Simão Lindoso de Souza
Vitória Araújo

Arbuscular mycorrhizae and family farming in the agreste of Paraíba

Simão Lindoso de Souza
Vitória Araújo

Arbuscular mycorrhizae and family farming in the agreste of Paraíba

Agricultural crops and their impact on and contribution to soil biological diversity

ScienciaScripts

Imprint
Any brand names and product names mentioned in this book are subject to trademark, brand or patent protection and are trademarks or registered trademarks of their respective holders. The use of brand names, product names, common names, trade names, product descriptions etc. even without a particular marking in this work is in no way to be construed to mean that such names may be regarded as unrestricted in respect of trademark and brand protection legislation and could thus be used by anyone.

Cover image: www.ingimage.com

This book is a translation from the original published under ISBN 978-3-330-76543-6.

Publisher:
Sciencia Scripts
is a trademark of
Dodo Books Indian Ocean Ltd. and OmniScriptum S.R.L publishing group

120 High Road, East Finchley, London, N2 9ED, United Kingdom
Str. Armeneasca 28/1, office 1, Chisinau MD-2012, Republic of Moldova, Europe
Managing Directors: Ieva Konstantinova, Victoria Ursu
info@omniscriptum.com

Printed at: see last page
ISBN: 978-620-3-47500-5

SUMMARY

The environment in which we live is rich in ecological relationships, in which we seek to benefit or exchange in order to reach a balance or stability. It's no different with plant species and soil microorganisms such as arbuscular mycorrhizal fungi (AMFs), which act as a complement to the host plant's root system and are able to increase nutrient absorption and thus provide greater growth and survival. The plant in turn provides the AMFs with energy from photosynthesis, which favours their growth and maintenance of their life cycle. Therefore, this association is an essential component for the development of sustainable agricultural and natural ecosystems. The aim of this study was to show the morphotypic occurrence of AMFs in family farm areas. The research took place in the municipality of Boa Vista, Agreste do Estado da Paraíba, where soil samples were collected from areas cultivated with maize, beans, fodder palm, a consortium of the three crops and managed Caatinga for chemical assessment and the presence of AMF spores. The spores were extracted using wet sieving and centrifugation in sucrose and morphotyped using morphological characteristics. The soils had medium acidity, low levels of organic matter and low phosphorus (P) availability. A total of 24 AMF morphotypes were found, with the consortium area having the highest number (19) and the maize monoculture area the lowest (14), although the latter showed the peculiarity of two exclusive morphotypes. The highest density of spores occurred in the maize area (985) and the lowest in the managed Caatinga area (360). Studying AMFs, emphasising their diversity, population and community, is a fundamental step for different approaches, both in understanding their symbiotic role in different ecosystems and in improving the production of agricultural ecosystems.

KEYWORDS: Ecology; Soil Microbiology; Soil Management.

CONTENTS

CHAPTER 1

INTRODUCTION

The environment in which we live is rich in ecological relationships. Our daily lives are made up of these relationships, whether with beings of the same species or with beings of different species. But we are always looking for a relationship of benefit or exchange in order to achieve balance or stability.

It's no different with plant species, as their relationship with micro-organisms is seen as an eminently relevant aspect, given that soil micro-organisms play a fundamental role in soil genesis and also act as nutrient regulators through the decomposition of organic matter and the cycling of elements. They therefore act as a source and drain of nutrients for plant growth (ANDREOLA, FERNANDES, 2007), especially when these micro-organisms are arbuscular mycorrhizal fungi (AMFs), which have a mycorrhizal association with around 80% of plant species (INVAM, 2015 - http: //invam.wvu.edu).

This type of plant-root fungus association is considered symbiotic and mutualistic. Symbiotic because the organisms co-exist in the same physical environment, root and soil, and mutualistic because, in general, both symbionts benefit from the association (BERBARA et al., 2006).

According to Colozzi Filho and Cardoso (2000), these fungi act as a complement to the host plant's root system, capable of increasing the absorption of P and other nutrients and thus also providing protection against pathogens. They can also increase plant tolerance to stressful situations such as low water content, salinity and the presence of heavy metals.

AMFs offer the plant a greater capacity for survival and growth, thus contributing to the success of plant populations. On the other hand, the plant provides the fungus with energy from photosynthesis, which guarantees its growth and maintenance.

The importance of AMFs for the sustainability of agricultural and natural systems can be understood by their wide occurrence in various terrestrial ecosystems, since they are the most abundant in agricultural soils (CARDOSO et al., 2010). This sustainability is linked to the beneficial effects of mycorrhizae on plant nutrition, especially in relation to the absorption of P, a non-renewable natural resource, as emphasised by Berbara et al. (2006). In this way, the increase in nutrient absorption is considered to be the primary factor of arbuscular mycorrhizae, as often the result of the presence of AMFs on plant growth is so significant that it can be replaced by the application of this nutrient (SAGGIN JÚNIOR, SILVA, 2005).

Another factor of great efficiency and benefit in increasing agricultural production, especially of legumes, is the symbiosis involving three beings: the AMFs, the plants and the nitrogen-fixing bacteria (N). Research involving AMFs and these nitrogen-fixing bacteria aims to increase production, minimise the use of chemical fertilisers and contribute to achieving a more sustainable and input-independent agricultural pattern (SIQUEIRA, MOREIRA, 1996).

AMFs also help to stabilise soil aggregates, bearing in mind that soil is an important property for controlling plant growth in arid and semi-arid environments by controlling soil-plant water status (RILLIG, 2004). Aggregation provides the soil with greater defence against erosive effects, which contributes to greater agricultural productivity.

Miranda and Miranda (1997) point out that the use of AMFs has been considered for 30 years as an alternative for reducing the use of fertilisers and pesticides in agriculture, due to their beneficial effects on the growth of plants of agronomic, forestry, horticultural and pastoral interest. However, the use of AMF inoculants in agriculture is still limited because there is no suitable technology for their production and also because there is little commercial interest in their production and commercialisation (SAGGIN JÚNIOR; LOVATO, 1999).

All in all, it can be emphasised that AMFs are essential elements in stabilising ecosystems due to the vital role they play, from their relationships with plant communities , influencing diversity and abundance, as well as the soil's aggregation capacity.

CHAPTER 2

OBJECTIVES

2.1 General objective

- To verify the occurrence of arbuscular mycorrhizal fungi in areas of different agricultural crops on family farms in the semi-arid region of Agreste Paraiba.

2.2 Specific objectives

* To diagnose the morphotypic occurrence of arbuscular mycorrhizal fungi in cultivated areas and in the managed Caatinga;
* Relate the presence and quantity of spores of AMF morphotypes to the type of agricultural crop;
* Evaluate the soil in terms of its chemical aspects and relate the influence of these factors on the occurrence of arbuscular mycorrhizal fungi in the areas studied.

CHAPTER 3

THEORETICAL FRAMEWORK

3.1 The AMFs

31.1 Classification and origin

AMFs are organisms belonging to the Kingdom Fungi. These fungi are grouped in the monophyletic Phylum Glomeromycota, Class Glomeromycetes. The group is currently made up of four orders (Glomerales, Diversisporales, Archaeosporales, Paraglomerales), 10 families and 18 genera and 215 described species. The current families are: Glomeraceae, Pacisporaceae, Acaulosporaceae, Diversisporaceae, Gigasporaceae, Claroideo- Glomeraceae, Paraglomeraceae, Archaeosporaceae and Ambisporaceae and the 18 genera are Funneliformis, Septoglomus, Glomus, Rhizophagus, Pacispora, Acaulospora, Diversispora, Redeckera, Gigaspora, Dentiscutata, Cetraspora, Racocetra, Scutellospora, Claroideoglomus, Paraglomus, Archaeospora, Ambispora and Geosiphon (REDECKER et al., 2013).

This classification is the one accepted by the website (http://invam.wvu.edu) - INVAM (International Culture Colletion of (Vesicular) Arbuscular Mycorrhizal Fungi) and this research is based on the information contained therein, as it is an internationally recognised website in the field and its database contains information on all aspects of FMAS.

However, until 2012 the classification was done differently, with the restriction that only the families of the order Glomerales were considered to form arbuscular mycorrhizae, in this case five families (Gigasporaceae, Glomeraceae, Acaulosporaceae, Paraglomaceae and Archaeosporaceae), of which there were a total of seven genera (Acaulospora, Archaeospora, Entrophospora, Glomus, Gigaspora, Paraglomus and Scutellospora).

The genetics of Glomeromycetes is still little known, but it was on the basis of studies in this area that this classification was based. Therefore, based on ribosomal RNA genes: 18S

(SSU), ITS1 - ITS2 - 5.8S (ITS), and/or 28S (LSU) (REDECKER et al., 2013). The origin of AMFs is dated to around 1200 to 1400 million years ago, long before the first land plants appeared, which was possibly 470 million years ago. However, it is not yet known when the evolutionary origin and symbiosis evolved (DE SOUZA et al., 2008); what is known is that studies indicate that they evolved from a common ancestor to the Basidiomycota and Ascomycota fungi group (JAMES et al., 2006). Another important piece of evidence is that phylogenetic analyses of nucleic acid sequences show that species of the genus Paraglomus represent the basal lineage of this group (REDECKER et al., 2000).

Glomeromycetes are fungi that form an obligatory association with plants, as they depend on them to complete their life cycle. The relationship is therefore seen as symbiotic and obligatory mutualistic. This association occurs through the interconnection of fungal hyphae with the roots of terrestrial, epiphytic and aquatic vascular plants, as well as with the rhizoids and stems of bryophytes and other basal plants (SOUZA et al, 2010). Root colonisation takes place through the penetration of hyphae into the cells of the plant root cortex, both between the cells and intracellularly (SAGGIN-JÚNIOR, SILVA, 2005).

The association of Glomeromycetes with plants is known as arbuscular mycorrhiza (AM), because they form a structure inside the plant root that is responsible for nutrient exchange, the arbuscle, formed by the interaction of hyphae and the plasmalemma of some cortex cells (BERBARA et al, 2006). This characteristic is shared by all members of the group, which is why the Glomeromycota phylum is considered monophyletic. Although these structures are extremely important in the process of exchanging metabolites between symbionts, it is not yet known which mechanisms control the functioning of the arbuscules (LAMBAIS, RAMOS, 2010).

Most representatives of the plant kingdom, from bryophytes to angiosperms, form mycorrhizae, but these are formed by combining different groups of host plants and different phyla of fungi. This reveals the existence of other types of mycorrhizae, which are differentiated by the morphological and anatomical characteristics of the roots and which are called ectomycorrhizae, ectendomycorrhizae, arbutoid mycorrhizae, ericoid mycorrhizae and orchidoid mycorrhizae.

Saggin-Júnior and Silva (2005) highlight each type of mycorrhiza: ectomycorrhizae are mycorrhizae formed by fungi of the phyla Basidiomycetes and Ascomycetes with Angiosperms and Gymnosperms, the hyphae do not penetrate the cells of the root cortex. They grow between and around the cells, forming Hartig's Net. They form a fungal mantle that surrounds the root, where the morphology of the root is visually modified. Ectendomycorrhizae are formed mainly by Basidiomycetes with Gymnosperms and Angiosperms, the hyphae penetrate inter- and intracellularly into the cells of the root cortex, forming tangled hyphae inside the cells, i.e. a thin fungal mantle that surrounds the root. Arbatoid mycorrhizae are formed by Basidiomycetes with plants of the genus Arbutus. They are similar to ectomycorrhizae in their root morphology, forming Hartig's Net and the epidermal cells are colonised, forming tangled hyphae inside, which disintegrate without affecting the cells, which can be colonised again. The ericoids are formed by Basidiomycetes and Ascomycetes with plants of the order Ericales and the family Monotropaceae, the surface of the roots has fewer hyphae than a typical ectomycorrhiza, which grow between the cells and invade them, forming tangled hyphae. Orchidoids are formed by Basidiomycetes with plants from the Orchidaceae family. During germination, orchids are dependent on an external carbon supply, which is provided by the association. In the adult plant, tangled hyphae and platoons are formed inside the cells of the root cortex (Figure 01).

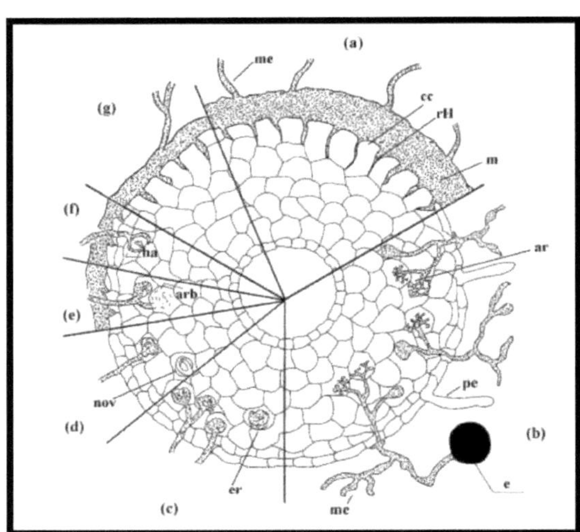

Figure 01 - Schematic of the types of mycorrhizae
a - ectomycorrhizae (ECM); b - arbuscular mycorrhizae (AM); c - ericoid mycorrhizae; d - mycorrhizae

orchidaceous mycorrhizae; **e** - arbutoid mycorrhizae; **f** - monotrepoid mycorrhizae; **ar - arbuscules**; **arb -** arbutoids; **cc -**
cortical cell; **e** - spore; er - coils; ha - haustoria; **m** - mantle; **me** - extraradicular mycelium; **nov** -novules; **rH** - Hartig's net.
Source: http://www.uc.pt/grasses/Divers_fungica/tipos_de_micorrizas (Azul, 2002, modified from Deacon, 1997).

3.1.2 Nutritional and ecological role

The most important role of AMFs is the acquisition of nutrients from the soil. They provide plants with greater absorption of water and nutrients, which are often fixed to soil structures and therefore difficult to access without the help of mycorrhizae. Due to its low mobility in the soil and the importance of phosphorus (P) for all plants, the absorption of this nutrient mediated by the association deserves special mention. Normally, P can only be absorbed by the root when phosphate ions are located in the region of immediate contact with the root surface (CARDOSO et al., 2010), hence the importance of the arbuscular mycorrhizal association, where the extraradicular hyphae act more efficiently, exploring a greater volume of soil that the root would not be able to reach.

Mycorrhizae can even contribute significantly to nutrient absorption, increasing absorption rates by up to 80% of P, 60% of Cu (Copper), 25% of N (Nitrogen), 25% of Zn (Zinc) and 10% of K (Potassium), as demonstrated by Marschner and Dell (1994).

Mohammad et al. (2004) point out that increased P nutrition in plants colonised by AMFs results, among other things, in increased growth and photosynthetic activity, an increase in the rate of carbohydrate transfer to the roots, a decrease in the pH of the rhizosphere due to colonisation, which can lead to increases in the solubility of P in the soil.

The different photoassimilates that reach the roots can act by regulating the activity of various enzymes, as well as the presence of sugars and amino acids (Berbara et al., 2006); this can favour greater growth of certain species of AMFs in various plant species and associations of AMFs.

MAs provide greater survival to plant communities by reducing their susceptibility to pathogens, a fact that can be partially explained by changes in plant physiology that activate defence mechanisms after mycorrhizal colonisation (FOLLI-PEREIRA et al., 2012). Reports on proteins related to pathogenesis, phytoalexin accumulation and cell wall lignification indicate the occurrence of systemic resistance in mycorrhised plants in locations far from the infection sites (SELOSSE et al., 2004).

AMFs are still important for maintaining the balance of riparian forests, since these areas suffer significant impacts and therefore need a long period of time to re-establish themselves. In this context, the balance can be accelerated by reforesting the forest using seedlings of plant species inoculated with AMFs. According to Colozzi-Filho and Nogueira (2007), AMFs have an efficient tolerance to drought and heat, and even under these stressful conditions, they are able to remove scarce nutrients from the soil and exchange them with the plant roots. In addition, they are able to maintain an ecological interaction with phytopathogens, helping to increase plant resistance to their action.

AMFs are also essential elements in the recovery of areas degraded by erosion, periodic flooding and mining, i.e. areas that have reduced the capacity of the soil to sustain the life of producer and consumer organisms, causing the ecosystem's function to decline. In this case, they represent a significant quantitative and qualitative element in ecosystems as they influence plant growth and adaptation to biotic and abiotic stresses (MOREIRA, SIQUEIRA, 2006), acting through various mechanisms, such as improving nutrition and favouring water relations.

Another important role of AMF colonisation is to reduce the negative effects of heavy metal pollution in the soil, such as aluminium (Al), and also to help with the water relationship with the plant when there is nutritional limitation, as the mycelial tangle helps to aggregate the soil. The outer mycelium of the AMF secretes glomalin, a hydrophobic glycoprotein that is highly stable in the soil and can remain there for 42 years until it is completely mineralised, so it is likely to have an impact on the construction of niches that will promote soil aggregation and structuring, with a consequent reduction in erosion processes (BERBARA et al., 2006).

3.1.3 Morphological structure of spores

Spores are one of the main propagules responsible for mycorrhizal colonisation, but there are two others, colonised root fragments and external mycelium, which functionally can contribute to the spread of AMFs in an ecosystem.

The morphology of AMF spores is essential if we are to be able to determine the identification of fungi in this group, at least at the taxonomic level of genus. Spores from this group are called glomerospores due to the characteristics that differentiate them from other groups (GOTO; MAIA, 2006).

The size, shape, colour, organisation and structure of the glomerospore wall, among others, are elements used to describe and identify AMFs, and the organisation of their walls is the component that provides the most information on the description of species. The walls of glomerospores are highly variable, unlike other groups of fungi, and can consist of up to four layers, which can be smooth or ornamented (INVAM, 2014).

According to Souza et al. (2010) glomerospore walls have been described since 1983 by Walker and until 1995 by Koske and Gemma. They described them, classifying them into ten types: a) Evanescent: ephemeral, single layer or laminated, with a gelatinous appearance; b) Unitary: Rigid, single smooth or ornamented layer, pigmented or hyaline; c) Laminated: Several thin, rigid laminae adhered together, smooth or ornamented, pigmented or hyaline; d) Expansive: single or laminated; e) Membranous: Thin, flexible wall, usually hyaline; f) Choriaceous: Flexible wall, 2.0 to 5.0 iim thick; g) Amorphous: Flexible but resistant wall; h) Chamfered: Flexible wall that fractures easily; i) Germinative: Single wall, concolorous to the laminar wall that precedes it, usually thin, with protuberances in the form of rounded papillae; j) Peridium: hyphal wrapping around isolated spores, in groups or sporocarps.

The colour of the glomerospores is a very important aspect in characterising the morphotypes, as they are diverse. They are hyaline, subhyaline, brown, light yellow, cream, orange, orange-brown, red-brown, brown to black, among others. However, the shape doesn't vary much, most of the time it's globose to subglobose, and can also be irregular, oval or pyriform. Their size can vary from 22 to 1050 iim (SOUZA et *al.,* 2010), showing that they are among the largest spores in the Fungi Kingdom.

Souza et al. (2010) point out that in the 1990s, researchers such as Morton and the Brazilians Franke and Sturmer carried out a comparative study of different genera of AMFs with the aim of defining the subcellular characters (walls) of the glomerospores based on ontogeny. The results of these studies revealed that glomerospore characters could be ordered hierarchically, according to their temporal and spatial origin; primary characters being the internal walls and germination structure; secondary, the distinct walls, highlighting colour, length and width; tertiary, the quantitative and qualitative aspects, such as thickness, colour, ornamentation and reaction to Melzer's reagent.

The ontogeny, i.e. the formation of glomerospores, has been described by five types to date. As Maia et al. (2010), the types are: a) glomoid forms the glomerospore in the terminal or intercalary portion of a hypha, it is more peculiar to the genus Glomus;b) gigasporoid forms the glomerospores at the apex of a bulbous sporogenous hypha, found in genera such as Gigaspora and Cetraspora; c) radial-glomoid forms terminally in the hypha, but in a radial manner, peculiar to the genus Glomus; d) aucalosporoid forms on the side of a hypha connected to a terminal spore sac, found in genera such as Aucaslospora and Ambispora; e) entrophospore forms the glomerospore inside the hypha connecting the spore sac, observed in genera such as Entrophospora. These types characterise taxonomic groups and are associated with different types of germination.

3.1.4 Physico-chemical characteristics of the soil x AMFs

Soil is an essential component for defining the biodiversity of a given area. To this end, it is made up of various aspects. Its main physical components are silt, sand and clay, and chemically it contains elements such as phosphorus (P), carbon (C), nitrogen (N) and hydrogen (H). However, it is the biological organisms present in the soil that particularly attract our attention. The organisms of the soil biota are bacteria, insects, nematodes, algae and fungi, which are fundamental to the renewal, maintenance and balance of the terrestrial ecosystem (PRIMAVESI, 2002). Therefore, soil is a resource that, together with its organisms, is fundamental to maintaining life and the balance of the biosphere (SOUZA et al., 2006).

The microorganisms present in the soil play an essential role in the productivity of agroecosystems and in the functioning of natural ecosystems (SIQUEIRA; FRANCO, 1988). Microbial activity in the soil can sometimes be harmful, but it is mainly beneficial through the action of AMFs, which, among other functions, contribute to the three-dimensional structuring of the soil made up of organic complexes, minerals (aggregates) and pores (RILLIG; MUMMEY, 2006).

AMFs contribute to soil quality in agricultural systems and play a vital role in sustainability, as they are possibly the most abundant soil fungi in such ecosystems (OLSSON et al., 1999). They are also capable of altering the physical and chemical characteristics of the substrate and therefore contribute to the formation and maintenance of soil structure, aggregating soil particles by means of extraradicular hyphae and their exudates and residues (FOLLI-PEREIRA et al., 2012).

The symbiosis between AMFs and plants is mutualistic in nature, as not only the plants are favoured, but also the fungi. The fungi acquire supplies and organic C compounds from the plants via photosynthetic fixation and return nutrients to the plants, especially in relation to P absorption (MOREIRA; SIQUEIRA, 2002).

This symbiotic relationship is possible because the fungi's intra- and extra-radicular hyphae are capable of absorbing mineral elements from the soil and transferring them to the roots, where they are absorbed and assimilated. In the intra-radicular environment, bidirectional exchange occurs mainly in the arbuscules, structures present in the cortex of the plant (BERBARA et al., 2006).

Several components are responsible for the good quality of soil nutrition. These include organic matter, the amount of phosphorus (P), carbon (C) and the pH value. Soil organic matter consists of all the organic material present in the soil, including litter, light fractions, microbial biomass, water-soluble organic substances and stabilised organic matter, known as humus (STEVENSON, 1994). The organic matter present in the soil stands out for its potential to indicate the quality of the soil, as it relates to other various components and exerts useful effects. Thus, Abbruzzini (2011, p. 20) states that organic matter has "a direct effect on water retention in the soil, aggregate formation, soil density, pH, buffer capacity, cation exchange capacity, mineralisation, absorption of pesticides and other agrochemicals, infiltration, aeration and microbial activity."

P is an element that has low availability in the soil because it has a high fixation capacity in the soil substrate. Because of this characteristic, plants find it difficult to absorb this nutrient, so there is a need to associate with AMFs to help them absorb it more easily. However, many times the plant can already be supplied with P and can dispense with the association with the fungus, as Costa et al. (2001) point out: the mechanism that regulates the plant-fungus association is related to the critical level of P for plant growth. In particular, Fernandes et al. (1987) point out that applying a small dose of P to soil with a high P deficiency will favour increased colonisation and sporulation of the fungi.

The absorption of N by AMFs is always related to their association with another microsymbiont, the nitrogen-fixing bacterium. When AMFs introduce P into the plant, they simultaneously contribute to greater nodulation of the legume by the rhizobium, since P is one of the indispensable elements for nodulation and N fixation (CARDOSO et al., 2010). AMFs can also act in the direct transfer of N from one plant to another through the communication of extraradicular mycelium or be responsible for increasing the utilisation of different forms of N by plants (CARRENHO, 2010). According to Sengik (2003), N has a faster effect on plant growth and promotes the development of the root system, thereby helping to improve the absorption of other nutrients from the soil.

C is an element that is richly stored in the soil and can be found mainly in the top layer between 0 (zero) and 30 cm, as well as at greater depths. The C present in the soil is influenced by management practices and edaphoclimatic factors, so inadequate soil management can take organic carbon into the atmosphere in the form of CO_2 through the mineralisation of organic matter (ABBRUZZINI, 2011). AMFs can be considered channels for the fixation of atmospheric C in the soil, via plants, as they have direct access to plant C sources. It is therefore estimated that, worldwide, they may be responsible for the annual drainage of five billion tonnes (5Gt) of C from soils (BAGO et al., 2000). This shows their great potential for sequestering CO_2 from the environment.

The pH value of the soil has an influential effect on the appearance and development of AMFs. Despite being found more predominantly in acidic soils, these fungi are also found in soils with a pH ranging from acidic to basic, in other words, according to each fungal species (BRANDÃO, 1992), or even each species adapts to a specific level of acidity or basicity in the soil. The predominance in acidic soils is probably due to the fact that there is less competition with other organisms and because the soil has less nutrients available for the plants. When the soil pH is between 6 and 7, the essential nutrients and micronutrients are absorbed more easily by the plants (MALAVOLTA, 2006 apud MARIANO, 2010).

The level of soil fertility is a factor that influences whether or not mycorrhizal association occurs. A high level of fertility tends to inhibit development, however, where there is low fertility there tends to be an accentuated association with the presence of greater colonisation and sporulation of the fungus (MOREIRA, SIQUEIRA, 2002). In this case we can say that the plant species are facultative symbionts, because when they have a high availability of nutritional supply, they can simply choose not to associate with the fungus. In this case, the inhibition is carried out by the plant through its own genetic mechanism (LAMBAIS, 2006 *apudBERBARA* et *al.,* 2006).

3.1.5 AMFs in natural ecosystems and agricultural crops

The occurrence of AMFs in various natural and agricultural ecosystems is determined and influenced by various biotic and abiotic factors, acting to alter or interfere with the process of their survival and colonisation in plants (CARRENHO et al., 2010).

In agricultural systems, in addition to these factors, soil management practices, crop management and cultural treatments, such as fertilisation and pesticide application, can lead to changes in the soil's physical, chemical and biological components (COELHO, 2008; CARRENHO et al., 2010). In this way, these factors can bring about quantitative and qualitative changes not only for the populations of AMFs present or for the cultivated plant species, but for both organisms involved in the symbiosis. It is therefore clear that arbuscular mycorrhizae have become multifunctional tools in agricultural systems, improving the physical, chemical and biological quality of the soil and, consequently, plant production and soil fertility.

According to Baumgartner et al. (2004), the community of AMFs and the degree of the fungus-plant association depends on how the vegetation cover is managed. In this way, it can be seen that AMFs will benefit plants if they are properly managed. Plants with a low mycorrhizal dependency have their productivity increased through the symbiotic relationship with the fungus only when the amount of phosphorus in the soil is low, while other species continue to benefit from high levels of phosphorus due to their high mycorrhizal dependency.

Stumer and Siqueira (2008) highlight the occurrence of almost 50% of the described species of AMFs in different natural and agricultural ecosystems in Brazil. Studies report that there is a greater occurrence of AMF species in unaltered ecosystems than in ecosystems managed with an agricultural system (SIQUEIRA et al, 1989; SIEVERDING, 1991).

The greater richness of AMFs in natural ecosystems is due to the greater abundance of plant species, which contrasts with agricultural systems with reduced plant diversity and soil and crop management, which is often done in the form of monoculture. In contrast, Miranda et al. (2005) and Siqueira et al. (1989), in studies, state that the highest number of spores and the highest rate of root colonisation are found in agricultural systems. However, this is due to the fungus' survival strategy in the face of reduced plant diversity.

Therefore, there is an increasing need to invest in strategies to improve and increase agricultural production, and many producers are already resorting to the use of AMF inoculums, which are an excellent alternative. Mergulhão et al. (2008) emphasise the importance of using arbuscular mycorrhizal fungi in crops of economic importance to Brazil, such as: pineapple, coffee, citrus, legumes, maize, papaya, tomatoes, as well as in the production of fruit tree seedlings and for reforestation, which can obtain up to a 30% increase in production, depending on the crop, when they use symbiosis.

3.1.6 AMFs in the Caatinga Biome

The Caatinga is an exclusively Brazilian biome that covers 844,453km^2, equivalent to 11% of the entire national territory (MINISTÉRIO DO MEIO AMBIENTE (MMA), 2014), mainly in the Northeast region. It is represented by tree and shrub vegetation that has adaptation mechanisms to withstand long periods of drought, such as deciduousness, i.e. the loss of leaves to avoid excess transpiration (MELO; RODRIGUEZ, 2004).

This biome has a very significant physiognomic variation, which has come to be called caatingas. This variation is favoured by factors ranging from regional and local climatic variations to floristic composition, topography, soil types and the impact of human activities. It's no wonder that Melo and Rodriguez (2004) group the Caatinga into four categories: 1) arboreal Caatingas, more or less dense, with or without cactaceae; 2) dense shrubby Caatingas, with few or many cactaceae and bromeliads; 3) more or less open Caatingas, with shrubs arranged in sparse clumps and a large number of cactaceae; and 4) sparse, low Caatingas.

One of the biggest challenges facing environmentalists today is the conservation of the Caatinga's biodiversity, as this biome suffers from a number of actions that have a significant negative impact. These include the devastation of the forest for agricultural and livestock activities, as well as for domestic and industrial supply, such as firewood, timber and charcoal. This extreme process of environmental degradation caused by the unsustainable use of natural resources has resulted in the decline or even extinction of animal and plant species and an increase in areas favourable to desertification. Unfortunately, the caatinga is the least protected natural vegetation in Brazil, with less than 2% of its entire area represented as conservation units (LEAL et al., 2003).

Microbiological diversity in the Caatinga is highly variable, mainly due to the high local variation in environmental factors such as levels of solar radiation, average annual temperature, relative humidity and periods of drought, which particularly influence the plant and animal life cycle (PRADO, 2003).

Studies show (GUSMÃO; MAIA, 2006; GUSMÃO et al., 2006) that there are fungi of various phyla in the Caatinga, totalling more than 800 species, and among the fungal phyla, there are representatives of the Glomeromycota phylum. In other studies (SOUZA et al., 2005; SILVA, 2003; ALBURQUERQUE, 2008; MERGULHÃO et al., 2009) the presence of glomeromycetes in native, preserved and impacted Caatinga is quite representative, due to the occurrence of more than twenty species in each study (21,24, 29 and 34 respectively). Therefore, in this context of caatingas, AMF are particularly important for their role in maintaining, balancing, increasing productivity and contributing to determining the floristic composition of terrestrial ecosystems.

Yano-Melo et al. (2003) highlighted the existence of 80 species of AMF in Brazil, only 1% of which were present in the Caatinga. However, more recent studies show records of 75 species of AMF in the Caatinga, distributed in natural, agricultural and impacted ecosystems (MAIA et al., 2010). This result shows a high level of diversity, given that we have just over 200 species recorded for Brazil, but despite the increase in diversity over the years, the scarcity of studies on the biodiversity of AMFs in this biome must still be taken into account.

In their study, Sturmer and Siqueira (2008) reported some very significant data for the Caatinga area. The occurrence of five species described in the Caatinga biome had not yet been recorded in Brazil, including one species of the genus Gisgaspora, three Glomus and one Scutellospora.

Checking the record of the diversity of AMF species by area in the Caatinga, it was observed that of the 75 recorded today, 27 occur in the three areas (agricultural systems, preserved areas and impacted areas) 15 occurred only in agricultural systems and 12 in preserved areas and two in impacted areas (MAIA et al., 2010).

CHAPTER 4

METHODOLOGY

4.1 Description of the study area

The study was carried out on two rural properties in the community called Caluête, located in the municipality of Boa Vista, in the semi-arid region of Paraiba.

Geographically, the municipality of Boa Vista belongs to the Mesoregion of Agreste Paraibano and the Microregion of Campina Grande. It is delimited by the coordinates 7°09'03.7" and 7°22'19.7" South latitude and 36°05'25.6" and 36°22'22.8" West longitude (Figure 02). It has a hot, dry climate with irregular rainfall distribution over short periods, a prolonged dry season and an average rainfall of 400 mm/year (SOUSA et al., 2008).

The vegetation is typical of the hypoxerophilous arbuscular Caatinga, especially the jurema, mandacaru, facheiro and catingueira, and the soils are shallow and stony (RODRIGUEZ, 2012). According to EMBRAPA (1999), the municipality of Boa Vista has five types of soil: Luvissolo Crômico Órtico vértico, Planossolo Nátrico Sálico típico, Neossolo Litólico Eutrófico, Afloramentos de rocha and some patches of Neossolo Flúvico Eutrófico.

Boa Vista is a new municipality, having been separated from Campina Grande only 22 years ago. It is also small and sparsely populated. This is shown by the Brazilian Institute of Geography and Statistics: its population is 6,227 inhabitants and its territorial area is 476,541 Km^2 (IBGE, 2010). However, this municipality stands out for its economic activities. The main ones are livestock farming (cattle, goats, sheep, pigs and poultry), agriculture (maize and beans), mineral extraction, such as bentonite, lime stone and rachinha stone, which demonstrates a very diverse rural area.

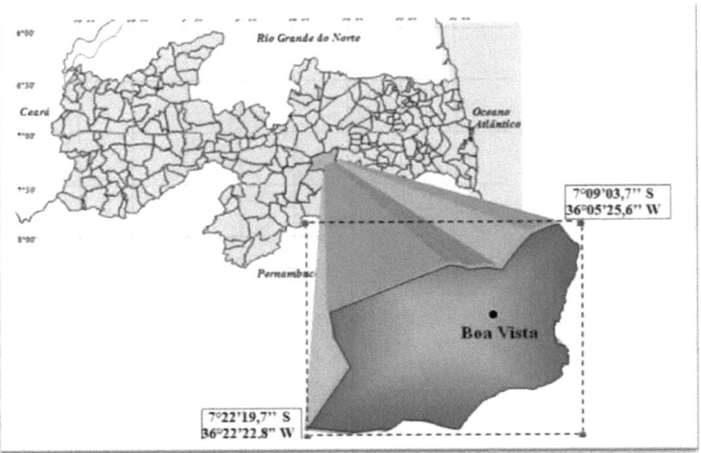

Figure 02 - Location of the Municipality of Boa Vista - PB
Source: SOUSA et al., 2008

4.2 Cultivation and management

On the properties studied, crops are grown mainly for family subsistence, but depending on the results of the harvest, they are also used for commercial purposes. Normally, crops are only planted in the rainy season (between January and June) and there is no use of irrigation. Most of the time, only maize and beans are grown. These two crops are usually intercropped, but can also be grown separately. It is also common for maize and beans to be intercropped with another type of crop, fodder palm. Fodder palm is a crop that serves as an alternative for farmers to feed their cattle, sheep and goats during long periods of drought.

The preparation of the area to be cultivated goes through different stages. Firstly, the Caatinga area is drilled and burnt, where all the plant species are cut down. A good number of the trees and shrubs that die naturally are used to make stakes, while others are used as firewood. What can't be used is thrown on the fire. After this stage, clearing is carried out, which consists of removing the stumps of the remaining trees. The soil in the cultivated area receives organic fertiliser from the manure of the cattle, sheep and goats, although this is done on an occasional basis. Insecticides are rarely used on the crops . The soil is managed mainly with animal-drawn ploughs and, occasionally, by tractors supplied by the municipality.

20

4.3 Soil sampling

Soil samples were taken in December 2013. The soil was sampled from each cultivation area using a Dutch auger at a depth of 20 cm. More than one sub-sample of rhizospheric soil was collected from each area, homogenised and used to make up a composite sample, which was air-dried and stored in duly identified plastic bags.

Five areas were studied, each represented as follows:

Table 01 - Five cultivation areas used to study the occurrence of AMFs

Areas	Cultivation
01	Maize (M)
02	Consortium of maize, beans and fodder palm (MFPF)
03	Managed Caatinga (CM)
04	Fodder palm (PF)
05	Beans (F)

Area 01(M) cultivated with maize has a total area of 8 (eight) hectares and is characterised by flat terrain with a gentle slope (Figure 03). Other crops have always been grown in this area, such as palm, beans and broad beans, often in intercropping. The fertiliser used has always been organic animal manure (cattle and sheep) and agrochemicals are applied sporadically to control caterpillars if they appear. Six sub-samples of soil were taken because the area has different aspects, such as high ground, high ground with turned over soil, low ground with stones and manure and the lowest part with crop stubble, in this case maize straw.

Area 02 (MFPF) was cultivated using a consortium of maize, beans and fodder palm (the maize and beans had already been harvested). It has a total area of 8 (eight) hectares and is characterised by a semi-inclined terrain, but with homogeneous soil throughout the area (Figure 03). In this area, the three crops were interspersed, i.e. one row of oil palm with maize and another of oil palm with beans. Three sub-samples were taken from for each crop combination.

Fodder palm is a crop that is more tolerant of pests, which is why it is an interesting strategy for family farming to grow it alongside other more susceptible crops. Crop rotation, fertilisation and the application of agrochemicals are carried out in the same way as reported in area 01.

Figure 03 - Agricultural cultivation areas 01: Maize (A and B) and 02: Maize, beans and fodder palm (C and D) in Boa Vista - PB.

Photo: Vitória Araújo (2013)

Area 03 (CM) is characterised by Caatinga, where we found representative species such as phacheiro, xique-xique, coroa-de-frade, cacti typical of this biome and semi-arid environments (Figure 04). Two sub-samples of soil were collected from this area, which had a darker colour due to the accumulation of organic matter of plant and animal origin, where the management of this area with some activities is clearly visible. Therefore, this area was called managed Caatinga because it is used for grazing cattle and sheep during the rainy season and, during the dry season, to harvest cacti for animal feed. Dry wood is also used to make fences and firewood for domestic use. However, this area is still in a good state of conservation, which has made it the reference area for relating the results obtained in the other areas.

Area 04 (PF), cultivated with fodder palm, has a total area of around 15 hectares and is characterised by a semi-flat terrain with yellowish soil (Figure 04). In this area we observed the presence of invasive plant species that were unable to withstand the dry season, as well as the presence of infestations of the carmine cochineal pest (Dactylopius opuntiae, Cockerell), which has been devastating this crop and putting it at risk of extinction in the semi-arid region. Two soil subsamples were collected from this area.

22

Area 05 (F), which had been cultivated with beans, has a total area of around 12 hectares and is characterised by flat terrain with reddish, stony soil in some places and lighter, sandy soil in others (Figure 04). In this bean-growing area, the presence of crop remains was not observed. Three soil subsamples were taken. Crop rotation, fertilisation and the use of agrochemicals are carried out in the same way as reported in area 01.

Figure 04 - Area 03: Managed Caatinga (E) and crop areas 04: Fodder palm (F and G) and 05: Beans (H) in Boa Vista - PB.

Photo: Vitória Araújo (2013)

4.3.1 Soil analyses

The soil samples were sent to the Irrigation and Salinity Laboratory of the Department of Agricultural Engineering at the Federal University of Campina Grande, and submitted for chemical analysis of pH in H_2O, Macronutrients, Assimilable Phosphorus and Organic Matter

.

4.4 Extraction, separation and counting of AMF spores

Three extractions of AMF spores were made from each composite sample. 50 ml of soil was used for the extraction procedure by wet sieving, using sieves with meshes of 500, 250 and 63 ^m in diameter, after which the sample obtained was centrifuged for three minutes at 3000 rpm (GERDEMANN, NICOLSON, 1963). The sample was then centrifuged for two minutes at 2000 rpm in a 50% sucrose solution (JENKINS, 1964).

The AMF spores were separated into morphotypes, i.e. specimens with similar morphological characteristics, taking into account colour, shape and size. Petri dishes and a stereoscopic microscope were used for this procedure. The morphotypes were then evaluated for direct counting in each morphotype group found, following the standards recommended by the International Culture Collection of (Vesicular) Arbuscular Mychorrhizal Fungi (INVAM, 2015 - http: //invam.wvu. edu).

Some spores representing the groups of morphotypes found were fixed on slides with PVLG (Polyvinyl lacto-glycerol) and observed under an optical microscope. The identification and description of AMF morphotypes was carried out by evaluating the morphological characteristics of the spores following the standards established by Schenck and Peréz (1987) and INVAM (2014) and using websites such as the Federal University of Lavras (http://www.dcs.ufla.br/micorriza/fungos_micorrzicos_arbusculares.html) and also (http://www.zor.zut.edu.pl/Glomeromycota/Taxonomy.html) as a tool.

CHAPTER 5

RESULTS AND DISCUSSION

5.1 Soil analysis

5.1.1 pH

The pH value is used to assess the conditions of a given soil, which can be acidic, basic or alkaline, which will influence plant productivity.

Most of the areas had a pH of medium acidity, with all values above five (Graph 01). Generally, acidic soils are characterised by the presence of toxic Al (Aluminium), which is harmful to plants and influences the development of the root system. However, from pH 5.5 there is no more toxic aluminium due to its precipitation in the form of aluminium oxide (MALAVOLTA, 1979), which is what can be seen in the five soil samples, all with pH values above 5.5 and zero Al (0.0).Area 03 (CM) had the highest acidity value, however, areas 01(M) and 04 (PF) had medium acidity. Area 05 (F) has weak acidity and area 02 (MFPF) had the most basic pH, i.e. weak alkalinity, as its value is below 7.8 (Tomé Jr, 1997).

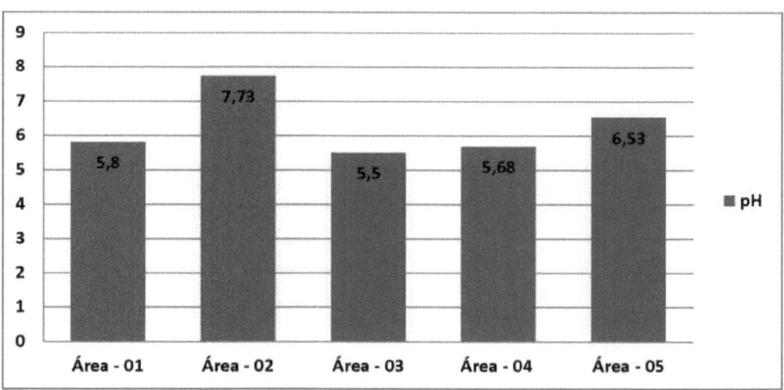

Graph 01 - Analysis of the pH ratio in H_2O **(1:2.5) in areas 01: Maize; 02: Maize, beans and fodder palm; 03: Managed Caatinga; 04: Fodder palm and 05: Beans in Boa Vista - PB.**

When the pH is acidic, as is the case in areas 01, 03 and 04, plants can suffer deficiencies due to the low availability of nutrients, which is due to the fixation of P (Phosphorus) by the elements Fe (Iron) and Al (Aluminium), forming insoluble compounds that are not available to plants. However, the greatest availability and absorption of nutrients by plants is found in soils with a pH range between 6.0 and 6.5, which is the most suitable for most crops (MALAVOLTA, 1979), which characterises the area 05 cultivated by beans, where it is known that this type of legume has its ideal pH between 5.5 and 6.5, so this area may have less favourable chemical conditions for the association of plants with AMFs.

Therefore, the ideal pH for growing maize is between 5.5 and 7.0, i.e. from weakly acidic soil to neutral pH, in which case area 01, cultivated by maize, showed a pH value within the ideal limits for its development.

If the pH values are below 4.5 or above 7.0, this indicates that the soil is in unfavourable conditions for plants, as is the case in area 02 cultivated with a consortium of corn, beans and fodder palm, with a pH value of 7.73, which causes a deficiency in P availability, This is what Sengik (2003) states when he says that the absorption of P by plants is directly influenced by the pH of the soil, and the more acidic or basic it is, the lower the absorption and availability of P will be. However, soils with a pH between 5.8 and 7.5 tend to be free of plant growth problems (SANTOS, 2008).

5.1.2 Macronutrients

Analysing the soil for its nutrients provides a picture of its fertility. Among the nutrients analysed, the Calcium ion was the predominant cation in the five areas, with values between 1.99 and 3.09 meq\100g. In a decreasing sequence, we see the distribution of Ca^2 ions+ $> Mg^{2+} > K^+ > Na^+$ in the five areas (Graph 02).

The Ca contents of the areas are considered average because they are above 2 meq\100g, with the exception of area 04 which had a value of 1.99 meq\100g. The availability of Ca to plants, as well as K and Mg, is affected by the amount of nutrient available in the soil and the degree of saturation in the exchange complex and the relationship with the other cations in the colloidal complex (SENGIK, 2003).

Since Ca and Mg values are used as soil fertility indicators, Orlando Filho et al. (1996)

indicate that there is an intimate relationship between these two nutrients in plant nutrition. This is due to the proximity of their chemical properties, such as ionic radius, degree of hydration and mobility, which leads to competition between adsorption sites in the soil and absorption in the roots, in which case the presence of one may jeopardise these two processes in the other.

Currently, there is an adequate ratio of Ca and Mg for plants, this ratio is based on Ca:Mg values between 4:1 and 8:1, however, in this study all five areas had a ratio below this proportion, in general an average of 2:1, which is not satisfactory for plants. However, Medeiros et al. (2008) explain that although this ratio of ideal proportions exists, it has not yet been established at what proportion of Ca and Mg in the cation exchange capacity (CEC) nutritional problems begin to occur in plants.

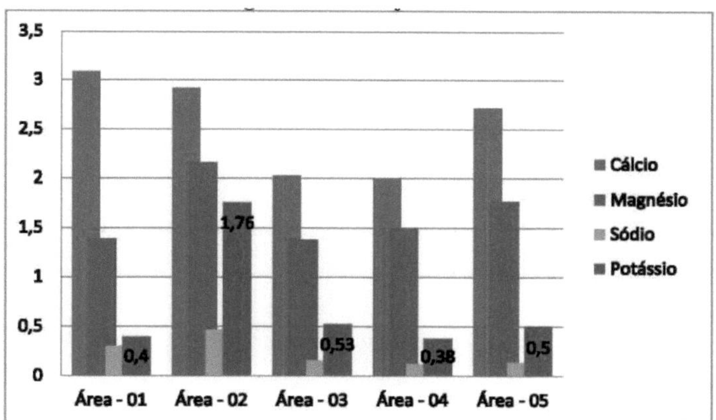

Graph 02 - Calcium (Ca), Magnesium (Mg), Sodium (Na) and Potassium (K) contents (meq/100g of soil) in areas 01 - Maize; 02 - Maize, beans and fodder palm; 03 - Managed Caatinga; 04 - Fodder palm and 05 - Beans in Boa Vista - PB

*equivalent measure

K is a nutrient that is not part of the composition of plants, however, it is absorbed in the form of a cation ion (K^+), regulating and participating in essential processes such as photosynthesis and water absorption; in forage plants its absorption is between 15 and 30 g of K\Kg of dry matter (SENGIK, 2003). In this study, the levels of both K^+ and Na^+ were well below those of Ca^{2+} and Mg^{2+}, which can be explained by the soil's low capacity to retain monovalent cations. K, on the other hand, is also found at a lower value because it is a very mobile element and is easily leached in soils with low CTC.

5.1.3 Organic Matter, Organic Carbon and Nitrogen

Knowing that the soil is rich or poor in organic matter can help you make appropriate recommendations for soil management. The organic matter values found in the soil samples from the five areas ranged from 11 to 19 g/dm³ of soil (Graph 03). Therefore, according to Ribeiro et al. (1999), the five areas had low levels of organic matter, since their quantitative interpretation considers values below 20g/dm³ of soil to be low, so in this case the medium level is only considered to be above 20.1 and the high level to be above 40.1g/dm3.of soil.However, the minimum value of soil organic matter for grain farming is 15g/dm³ (1.5%) (BARROS, 2015), which in this case, only area 02 with a crop consortium is above the limits, with a value of 19g/dm3, which is far below the limits of the areas cultivated with grains, with a value of only 13.3g/dm3.

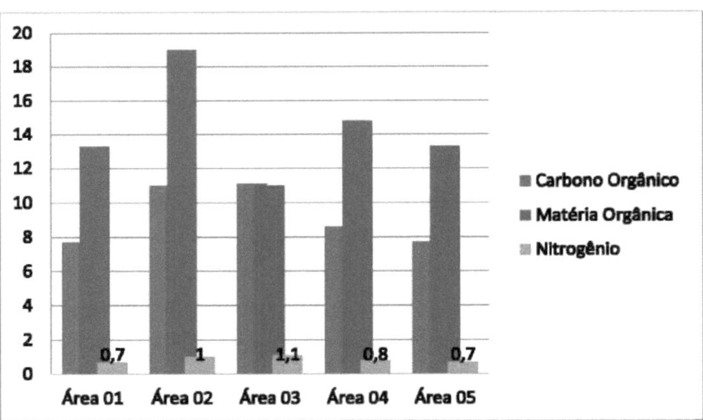

Graph 03 - Amount of Organic Carbon, Organic Matter and Nitrogen (g/dm³ of soil) in areas 01 - Maize; 02 - Maize, beans and fodder palm; 03 - Managed Caatinga; 04 - Fodder palm and 05 - Beans in Boa Vista - PB

Low organic matter content in the soil qualitatively indicates that there may be a deficiency of some macronutrients, such as N and S, as well as micronutrients, low cation exchange capacity, i.e. low buffering power and high base leaching power (Ca, Mg and K). Factors that contribute to low organic matter levels are low rainfall, low humidity, little vegetation, unfavourable biological conditions and soil management practices. Soil aggregation is the physical characteristic most compromised by low organic matter content, because by affecting aggregation, it also indirectly affects other physical characteristics of paramount importance for productivity, such as density, porosity, aeration, water retention and infiltration capacity (STEVENSON, 1994).

N, along with K and P, are considered primary macronutrients, i.e. the ones that plants need the most. N is responsible for increasing the plant's leaf area, which increases the efficiency of the photosynthetic rate and yield components and, consequently, grain yield (BARROS, 2015).

Most of this element is found in the NO_3^- form and is closely related to the amount of organic matter in the soil, as over 90% of the N available in the soil is found in organic form. The N values found in the five areas ranged from 0.7 to 1.1 g/dm^3, with area 03 having the highest level, while areas 01 and 05 had the lowest values. Low N values affect plant growth and the absorption of other nutrients.

The amount of organic carbon found varied between 7.7 and 11.7g/dm^3 in the five areas. These levels are considered low because all of them had values of less than 11.7g/dm^3 (RIBEIRO et *al.*, 1999), a result that may be related to the fact that the AMFs did not complete their life cycle due to unfavourable environmental conditions, in this case, water stress did not allow the plants to develop properly.

Low N and organic C values in this case are due to the use of a conventional planting system which, in turn, prepares the soil through ploughing and harrowing, turning the soil and attaching plant residues, which leaves the topsoil unprotected, subject to erosion and losses of C and N (BRAGA, 2016). Therefore, adhering to conservation management, such as crop rotation, would be a good alternative for increasing N values, since it favours the increase and recovery of organic matter, as it reduces the amount of disturbance and residues that cover the soil surface.

5.1.4 Assimilable phosphorus

Assimilable phosphorus is the part of P that is diluted in the soil solution, i.e. that which is easily absorbed by plants. The natural source of P in the soil is organic matter and it is absorbed by plant roots in the form of the orthophosphate ion ($H_2PO_4^-$) (SENGIK, 2003).The study showed that the lowest content of assimilable P found was 3.45 and the highest was 4.5 mg/100g, so the five areas have an average of 3.79 mg\100g, with area 02 having a content above this average (4.5 mg\100g), which is due to the fact that this area has a greater amount of organic matter and is a three-crop consortium area. However, these values are still considered low.

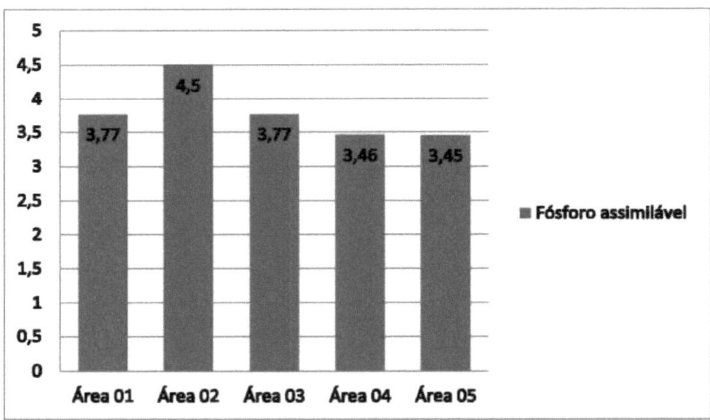

Graph 04 - Assimilable phosphorus (P) content (mg/100g) in areas 01 - Maize; 02 - Maize, beans and fodder palm; 03 - Managed Caatinga; 04 - Fodder palm and 05 - Beans in Boa Vista - PB

The result found contradicts Sengik's (2003) statement that pH in the range of 6.0 and 6.5 provides greater availability of P due to the minimal reaction and fixation of P in this range, but the only area that fell within this range was area 05, which had the lowest assimilable P content.

However, this low P content found in all the areas contributes to the establishment of arbuscular mycorrhiza, since low levels of P increase the colonisation of AMFs, whereas in soils with high levels of P, root colonisation is reduced or may even be absent, i.e. the symbiosis may lose its mutualistic character and give way to a parasitic relationship because the symbiotic balance is unfavourable to the plant, since there is a high cost of photoassimilates to maintain the fungus-root relationship (SMITH; READ, 1997).

5.2 Occurrence of AMFs in agricultural crops and in the managed Caatinga

5.2.1 Occurrence and density of AMF spores

The results show the presence of a significant morphotypic variety of AMFs in all five areas studied (Figures 05, 06 and 07), with a total of 24 AMF morphotypes (Table 01).

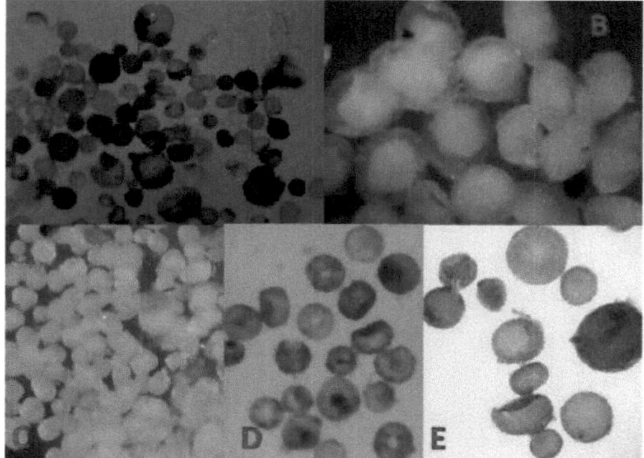

Figure 05 - Morphotypes of AMFs found in area 01 - Maize (A, B and C) and area 02 - Maize, beans and fodder palm (D and E) in Boa Vista - PB.
Photo: Vitória Araújo (2014)

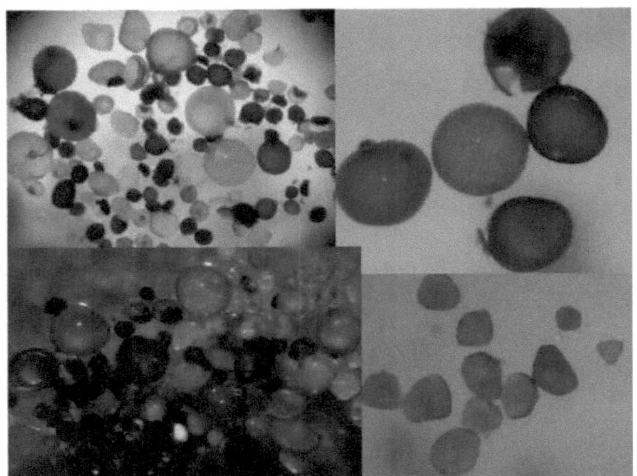

Figure 06 - Morphotypes of AMFs found in area 03 - managed Caatinga in Boa Vista - PB.
Photo: Vitória Araújo (2014)

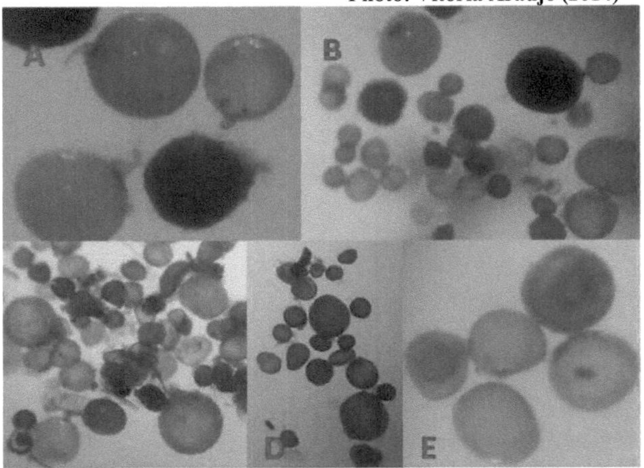

Figure 07 - Morphotypes of AMFs found in area 04 - Fodder palm (A and B) and area 05 - Beans (C, D and E) in Boa Vista - PB.
Photo: Vitória Araújo (2014)

The area with the greatest morphotypic variation was area 04 (MPF) with 19 morphotypes; in second place was area 05 (F) with 18 morphotypes, followed by areas 02 (MFPF) and 03 (CM) with 17 each. Area 01 (M), on the other hand, had only 14 morphotypes, although it stood out for having two peculiar morphotypes (4 and 14), which none of the other areas had. It was also found that ten (5, 6, 7, 8, 9, 10, 11, 12, 13 and 24) of the 24 morphotypes occur in all five areas, which are therefore called generalist morphotypes.

Miranda (2008) points out that, in general, natural ecosystems have a greater diversity of AMFs than agricultural ecosystems, but the latter have a greater number of spores, favouring the dominance of genera or species of AMFs with greater adaptability to environmental variations

In general, it can be seen that the five areas had a significant level of occurrence and that even though areas 04 and 05 had the most morphotypes, most of the morphotypes had few spores.

The highest spore density was found in area 01, which was cultivated with maize, as it had an average of 985 spores/50 ml of soil. A study carried out in family farming areas in Minas Gerais (FERNANDES et al., 2012) showed that the area cultivated with maize also had a higher number of spores (397/50 ml soil). Kelly et al. (2005) corroborate this result, as they state that plants from the Poaceae family, to which maize belongs, act as spore multipliers.

Area 03 - managed Caatinga was the one with the lowest density of spores/extraction, only 360, showing what Miranda et al. (2005) proposed, that in general, a greater number of spores occur in cultivated areas than in soils with natural vegetation, and which confirms other studies, where the number of glomerospores in Caatinga regions is always low (SOUZA et *al.,* 2003; BORBA; AMORIM 2007; MELLO et al., 2012)Taking into account the number of spores per morphotype, morphotype 9 had the highest number, with a total of 932 spores present in the five areas. Morphotypes 3, 4, 6 and 8 also had a significant number, which may indicate that these species are competitive with each other and that they are well adapted to these types of crops, so they use the sporulation strategy until favourable conditions are available for them to complete their life cycle.

Table 01 - Classification of morphotypes and spore density found in the agricultural cultivation and managed Caatinga areas

MORPHOTYPES	NUMBER OF SPORES					
	AREA 01 M	AREA 02 MFPF	AREA 03 CM	AREA 04 PF	AREA 05 F	TOTAL

1	0	0	13	1	1	15
2	5	5	0	0	1	11
3	130	0	23	0	70	223
4	232	0	0	0	0	232
5	1	16	1	4	6	28
6	30	77	31	40	52	230
7	11	11	3	1	4	30
8	61	91	14	13	45	224
9	443	142	103	113	131	932
10	23	14	24	68	41	170
11	4	3	6	12	11	36
12	7	3	8	52	7	77
13	24	16	40	10	23	113
14	11	0	0	0	0	11
15	0	68	0	49	0	117
16	0	4	0	3	7	14
17	0	4	10	1	6	21
18	0	21	0	1	0	22
19	0	4	0	0	1	5
20	0	1	3	1	1	6
21	0	0	1	2	0	3
22	0	0	61	1	1	63
23	0	0	1	1	0	2
24	3	4	18	26	22	73
TOTAL	985	484	360	399	430	2658

5.2.2 Description of AMF glomerospore morphotypes

Description of the 24 morphotypes occurring in the five areas.

Morphotype 1:

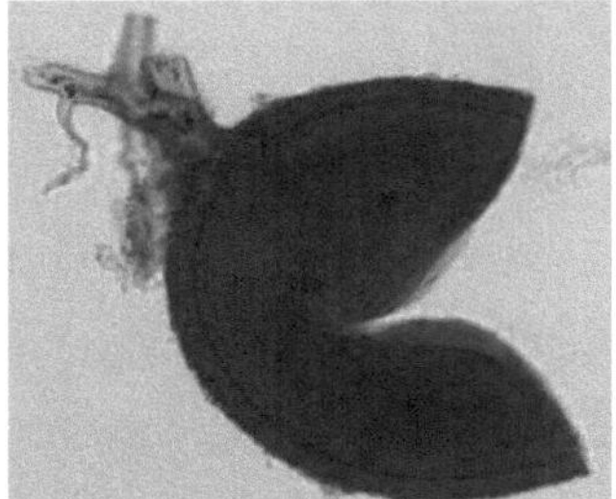

The glomerospore of this morphotype is characterised by its brownish orange to dark brown colour, small to medium size, globose to subglobose, light brown to brownish hyphae with branching and an opening pore. The wall has at least two layers, one of which is quite thick.

This morphotype resembles species of the genus Glomus, such as G. australe and G.

constrictium. The former, as described by Freitas and Carrenho (2013), has spores with very similar characteristics: reddish-brown in colour, globose to subglobose, smooth and opaque surface, robust appearance like a chicken thigh. Two-layered wall: C1 mucilaginous light brown and C2 unitary, smooth, firm with a reddish-brown colour. The hyphae are straight and some are branched and have an opening pore.

However, G. *constrictium is* described on the website (http://www.zor.zut.edu.pl/) as an isolated spore with an orange-brown to dark brown colour, globose to subglobose shape, sometimes ovoid and presenting sporogenous hyphae. Its wall is made up of two layers, C1 is evanescent, hyaline to pale yellow, normally present as an intact structure or more or less deteriorated, depending on the age of the spores and the microbiological activity of the soil and C2 is laminated, soft orange to dark brown and the hyphae are orange-brown to dark brown and can be straight or curved; generally marked by a narrowing at the base of the spore.

Morphotype 2:

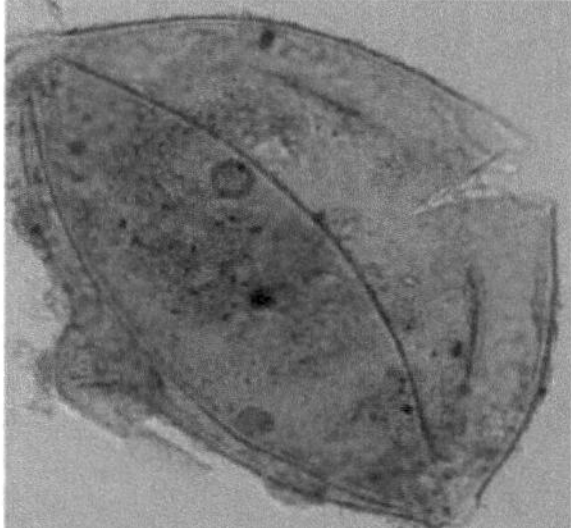

The glomerospore of this morphotype is characterised by its light yellow to orange colour, globose to subglobose shape, small to medium size, wall with at least two layers, development from sporogenous hyphae with a wall that continues that of the spore and is similar in colour. The layer appears to be mucilaginous in which organic particles are adhered.

This morphotype resembles species of the genus Glomus in that all spores of this genus have only one wall comprising at least two layers, of which the structural layer is made up of several sub-layers (laminae); the outer layer or the layers adhering to the laminate layer often crumble with age (http://www.zor.zut.edu.pl).

Glomus has glomoid spores, produced on or near the soil surface, formed singly, in groups or sporocarps, usually with partial peridium; colour varies from light yellow, yellow-orange, pale to dark brown and yellow-brown, globose, subglobose, ellipsoid and irregular shapes (http://invam.wvu.edu).

Freitas and Carrenho (2013) describe the spores of the genus *Glomus* as formed terminally or interspersed in a fertile, cylindrical or enlarged sporogenous hypha, with a multi-layered wall, from two to four; hyphae of a similar colour or slightly lighter than the spore; pore open or occluded by a septum, all spores are produced isolated, aggregated or in sporocarps.

Morphotype 3:

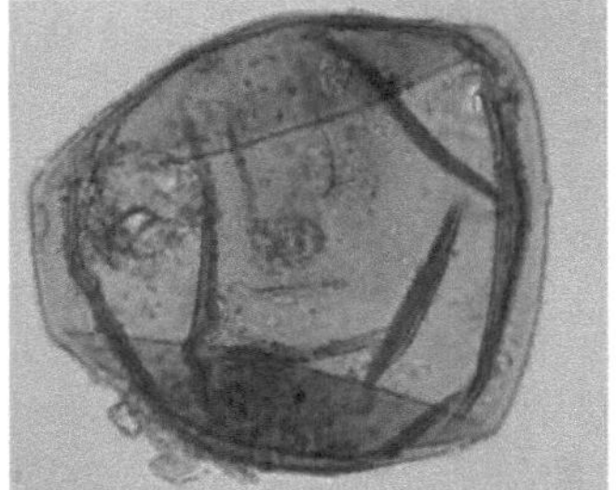

The glomerospore of this morphotype is characterised by its translucent yellow-orange colour, its globose to subglobose shape and most of the time it has a crumpled or shrunken appearance, its size varies from small to large, its wall has at least two layers and it has no ornamentation. In no representative was it possible to observe the presence of sporogenous hyphae.

This morphotype resembles the spore *of Pacispora scintillans.* It is described as solitary, hyaline to orange in colour, globose to subglobose in shape, rarely ovoid and with a subtended hypha; wall composed of three layers: C1 is permanent, rigid, hyaline to orange, ornamented mainly with warts, rarely, with furrows, tightly adherent to C2; however, separation from C2 occasionally occurs in vigorously crushed spores; C2 is laminated, smooth, hyaline, and C3 is flexible to semi-flexible, hyaline, usually tightly adherent to C2 (http://www.zor.zut.edu.pl). Oehl and Sieverding (2004) describe the *P. scintillans* spore as globose to subglobose to ellipsoid, never forming sporocarps, subhyaline to white in colour, three layers form the thin, subhyaline wall.

Morphotype 4:

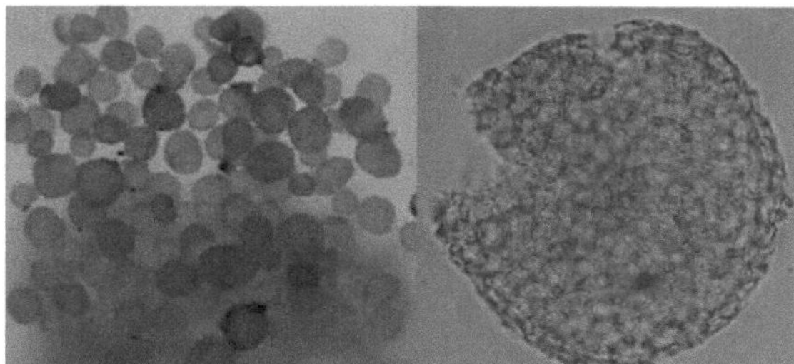

This morphotype is characterised by a light yellow to cream colour, but it differs from the others in that it has a spongy appearance, a subglobose to ovoid shape and varies in size from small to large. However, it was not possible to observe the presence of hyphae, nor was it safe to describe the minimum number of layers. No spores with a similar appearance were identified in the specific literature.

Morphotype 5:

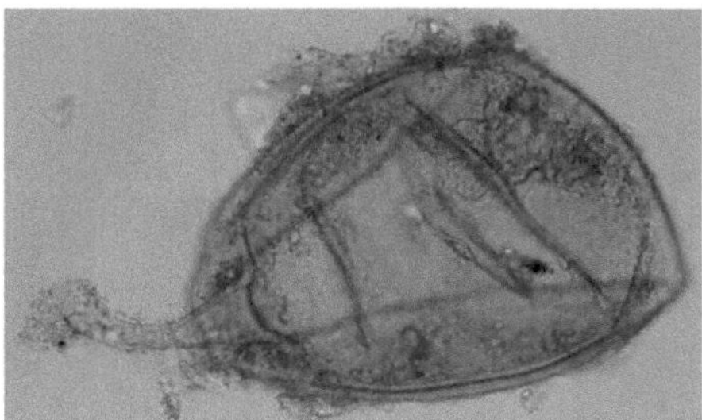

The glomerospore of this morphotype is characterised by a light yellow colour with a translucent appearance, a globose to subglossy shape, a small to large size, sporogenous hyphae with a tapered shape of the same colour or hyaline, and a curved septum. The wall has at least two layers.

Morphotype similar to species of the genus Funneliformis which is described by Freitas and Carrenho (2013) as follows: spores formed terminally in a fertile hypha, generally tapered, wall with two to three layers, sporogenous hypha similar in colour or slightly lighter than the spore, pore occluded by a curved septum, far from the base of the spore.

Funneliformis has pigmented spores formed singly in the soil or in groups of one to approximately twenty spores surrounded by an entire or partial mycelial mantle, often with a funnel-shaped spore base. The spore wall is usually made up of two or three layers, the outer one being hyaline. The spore is usually occluded by a distal septum at the base (http://invam.wvu.edu).

Morphotype 6:

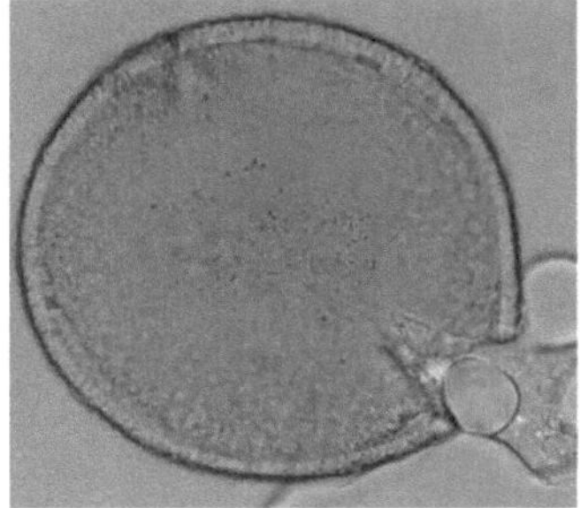

The glomerospore of this morphotype is characterised by being light yellow in colour with a translucent appearance, globular in shape, varying in size from small to large, and an outer wall with a layer of medium thickness.

This morphotype is similar to the species Glomus pansihalos, whose spore is described at (http://www.zor.zut.edu.pl/Glomeromycota/Glomus%20pansihalos.htm) as unique in the soil, pale to dark yellow in colour, globose to subglobose in shape and with the presence of sporogenous hyphae, a wall composed of three layers, where C1 is flexible, hyaline and expands in PVLG into a radiant halo, similar to a crown that disappears over time; C2 is laminated, pale to dark yellow in colour, decorated with evenly distributed hemispherical warts, and C3 is flexible to semi-rigid, hyaline to pale yellow, generally easy to separate from C2 in crushed spores.

Berch and Koske (1986) describe them as spores that vary from globose, subglobose, ellipsoid to irregular, with a yellowish-brown to dark brownish-orange colour in transmitted light, in incident light the surface of the spore is opaque white and scaly exposing more of the inner walls, supported on a single hypha; the wall composed of three inseparable layers.

Globose, ellipsoid spore, irregular in shape, yellow-brown to dark brown in colour; wall consists of three layers, where the outer layer (C1) is hyaline, and shows swelling and expansion in lactic acid or PVLG and forming hyaline to pale yellow columns that radiate outwards from the surface of the spore; C2 is rigid, yellow-orange to orange-brown, covered with spaced warts, and the innermost layer (C3) is yellow-orange; it has single, straight hyphae, sometimes recurved at the point of attachment, pale yellow in colour, description from the website (http://invam.wvu.edu).

Morphotype 7:

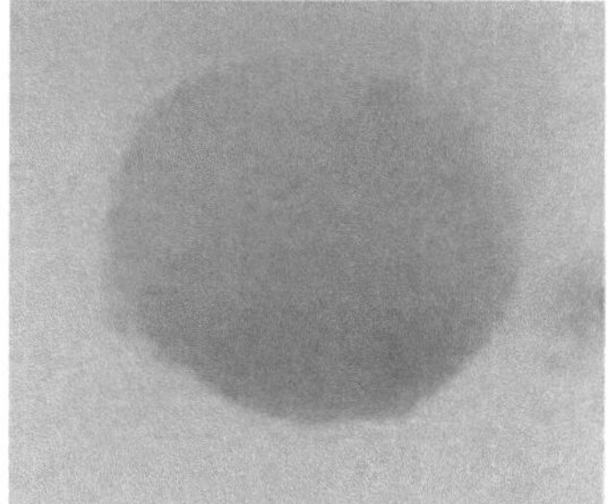

The glomerospore of this morphotype is characterised by its bright white colour, globose shape and size, which varies from small to large; no sporogenous hyphae were seen under the stereomicroscope. It was not possible to make a slide to better visualise the structures. However, this morphotype resembles spores from the genus *Diversispora*, since spores from this genus can be white in colour.

Morphotype 8:

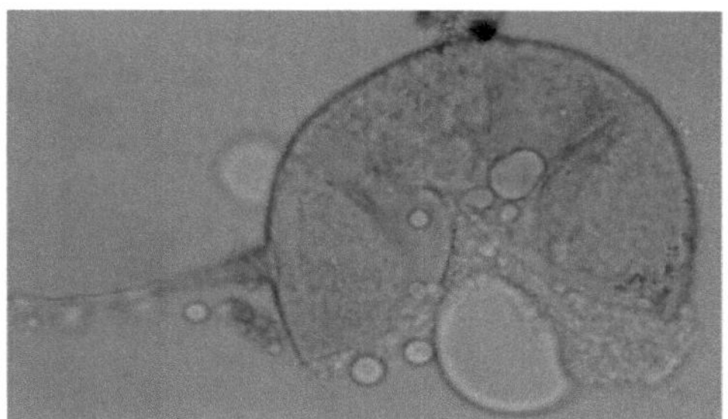

The glomerospore of this morphotype is characterised by being hyaline to subhyaline with luminescence due to shiny dots inside, which is probably a lipid reserve. Size from small to large, globose to subglobose in shape, with sporogenous hyphae, where this appears thin and similarly coloured, in some morphotypes it was not possible to see it. The wall of the morphotype has at least two layers, and the outer layer of the spore is continuous with the wall of the hypha. In some glomerospores, organic matter can be seen attached to the body or even to the hypha.

According to Freitas and Carrenho (2013), this morphotype is similar to a species from the genus Claroideoglomus, as its spore has hyaline sporogenous hyphae, a spore wall with two to four layers, the spore wall is continuous with the hypha wall, and the pore is occluded by the flexible layer. The website (http://invam. wvu.edu) describes the genus *Claroideoglomus* as having spore development from the blastic expansion of the hypha. The hyphae that underlie the spore differentiate and synthesise the same components present in the layers that are found in the spore wall. In some species of mature spores, the hypha is so thin that it is difficult to see, often because it has already separated from the spore, so it goes unnoticed.

Morphotype 9:

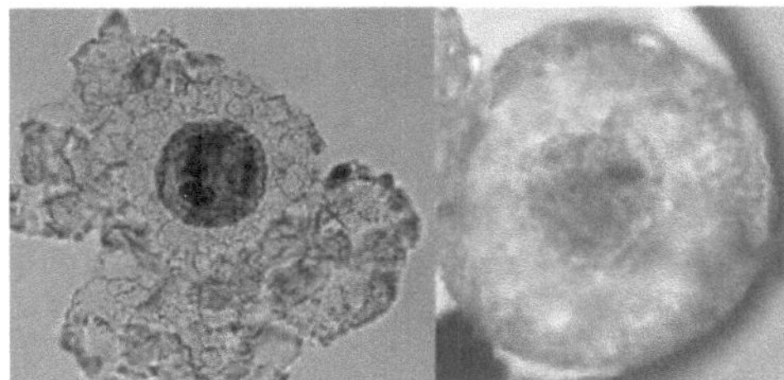

It is characterised by a hyaline colour, a globular shape with concavity, inside which there is often a substance that is probably a lipid reserve, the size varies from small to large, although small size prevails. The wall is fragile and appears to have only one layer. No spores with a similar appearance have been identified in the literature.

Morphotype 10:

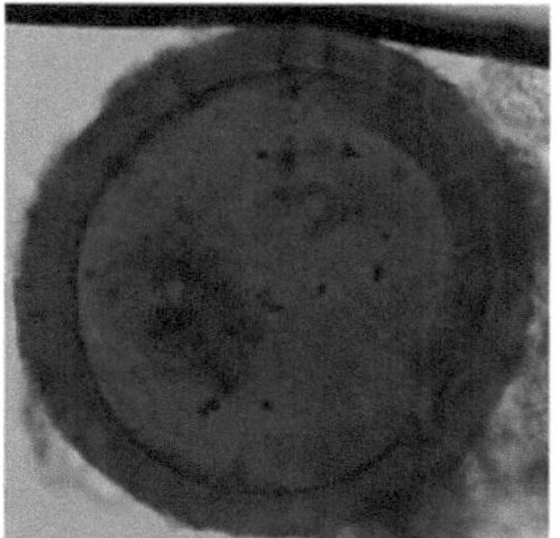

The glomerospore of this morphotype is characterised by its brownish-orange colour, globose to subglobose shape, size varies from small to large, the wall has several layers, one of which is quite thick. There are also hyaline to subhyaline sporogenous hyphae, which are often difficult to see. Pore occluded by one of the layers.

This morphotype resembles spores from the genus Acaulospora, as spores from this genus can range in colour from hyaline to reddish-orange.

In this case, it is very close to A. tuberculata, since it has spores with a red-orange to red-brown colour, a globose to subglobose shape, a three-layered wall, where C1 is hyaline, C2 is a layer that thickens initially with the formation of reddish-brown sub-layers, followed by polygonal synthesis, spines or tubercles, therefore with a thickness that reaches an average of 9.24^m, and C3 is yellow-brown in colour and appears discreetly due to the separation of the spore wall, which appears to be sub-layers of C2 (http://invam.wvu.edu).

Morphotype 11:

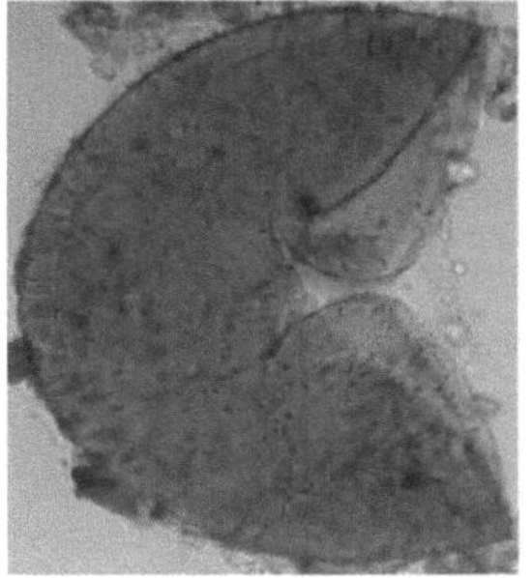

The glomerospore of this morphotype is characterised by its orange colour, globose to subglobose shape, medium to large size, multi-layered wall, the outer layer being thicker than the others, and no hyphae.

This morphotype resembles spores of the genus Acaulospora because it has a similar colour and shape to some species, the structure of the wall and the fact that it does not have hyphae is due to the fact that this genus develops from a sporiferous sacculus.

Morphotype 12:

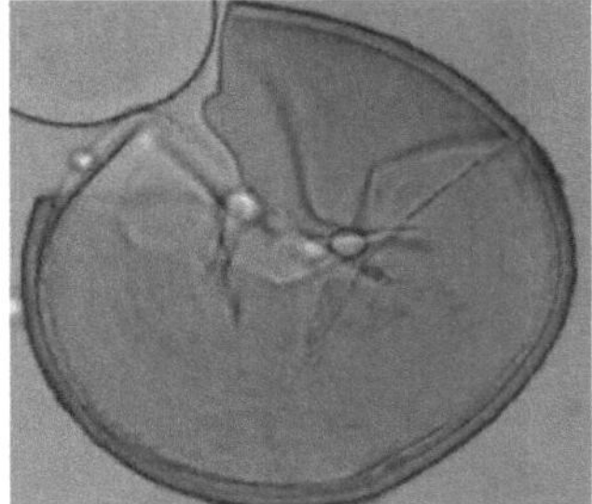

The glomerospore of this morphotype is characterised by a light orange colour with a translucent appearance, which has been observed in a cluster of spores or even forming sporocarps. An average of 20 to 30 spores arranged in a tangle of hyphae. The spores were only small and globose in shape. The wall has at least two layers.

This morphotype resembles species of the genus Sclerocystis, since Freitas and Carrenho (2013) describe this genus with spores formed at the apex of a sporogenous hypha, belonging to a branched web that gives rise to many peripheral spores, arranged side by side, spore wall consisting of two layers and sporocarps with or without peridium. Forming glomoid spores in sporocarps with a peridium, radiating from a central sterile plexus of the mycelium (SCHUBLER, WALKER, 2010).

Morphotype 13:

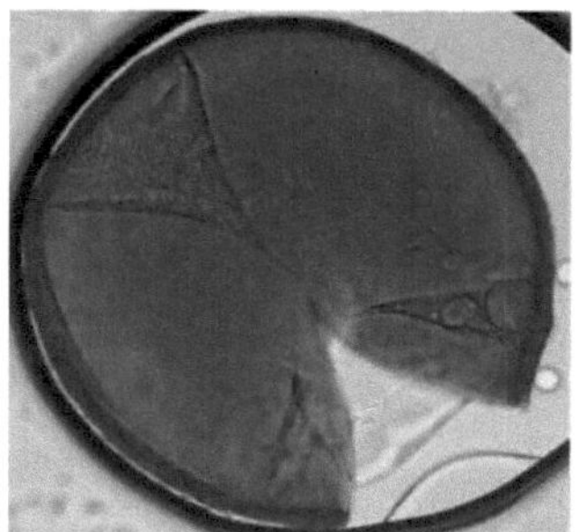

The glomerospore of this morphotype is characterised by its light to dark brown colour, globose to subglobose shape, size varying from small to large, smooth and resistant appearance, wall with at least three layers, the outermost of which is medium thick. There is also the presence of a scar.

This morphotype resembles spores of the genus Acaulospora, which can be subhyaline in colour, pale light yellow, orange-red, red-brown to orange-brown, globose to subglobose in shape, sometimes irregular, with two to three wall layers, two germinal walls with two layers each, an ovoid scar indicating the region of contact between spores and the neck of the sacculus during spore synthesis, and can be ornamented or not (http://www.dcs.ufla.br/; http://invam.wvu.edu).

Morphotype 14:

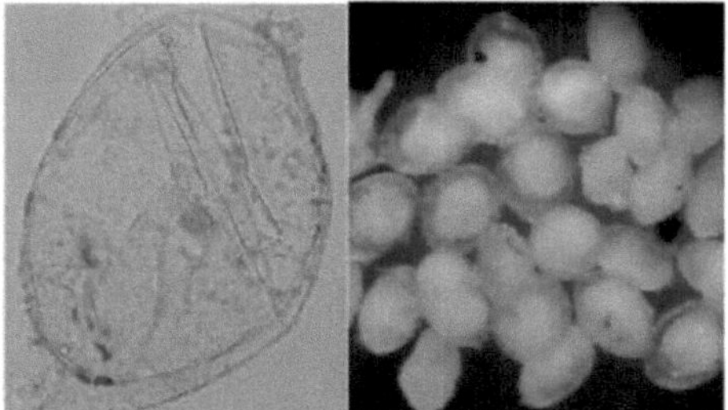

The glomerospore of this morphotype is characterised by being hyaline with a luminescent appearance, small to medium size, subglobose or ovoid shape, as well as being quite fragile, breaking easily when handled. The outer layer is hyaline and the nucleus or innermost layer is whitish.

This morphotype is similar to the Archaeospora schenckii species due to the fact that the spores of this species are colourless, have a three-layered wall, the innermost of which is completely detached, which resembles an endospore (FREITAS, CARRENHO, 2013). The spore of the A. schenckii species is described as completely hyaline (foamy white) when mature, mainly globose in shape, but also ellipsoid and ovoid; wall composed of three hyaline layers, which exhibit a certain flexibility (http://invam.wvu.edu).

Morphotype 15:

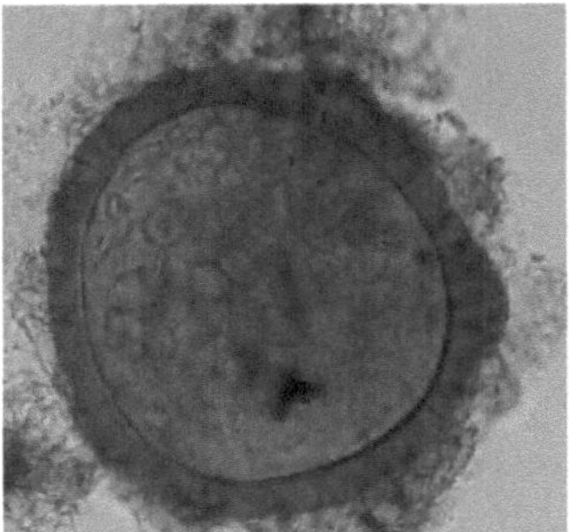

The glomerospore of this morphotype is characterised by its translucent yellow-orange colour, its globose to subglobose shape and its size, which varies from small to large. It has at least three layers, the outermost of which is evanescent and hyaline; another layer is quite thick. Its innermost layer apparently contains lipid droplets.
There is also a pore occluded by the flexible layer.

This morphotype resembles spores of the *Rhizophagus fasciculatus* species as it is pale yellow to brownish yellow in colour and globose to subglobose in shape. The wall is made up of three layers, where C1 is the outer, hyaline layer; C2 is made up of thin sub-layers or adhered laminates with a consistent thickness and C3 is thin and flexible and comprises the innermost layer of the hypha (http: //invam. wvu.edu).

Morphotype 16:

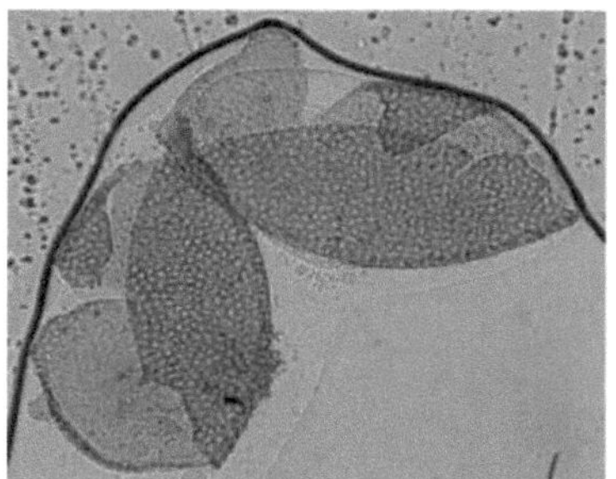

The glomerospore of this morphotype is characterised by its pale light yellow colour, its globose to subglobose shape and its size, which ranges from small to medium. Its wall is ornamented and visibly has at least two layers.

This morphotype is similar to the species *Acaulospora scrobiculata*, since the spore of this species has several characteristics in common. Its description on the website http://www.dcs.ufla.br/micorriza/fma is as follows: yellow colour (straw-coloured), globose to subglobose shape, wall composed of three layers. C1 is hyaline, evanescent, adhered to C2, continuous with the wall of the neck of the sporiferous sacculus, often completely lost in mature spores. C2 consists of yellowish white to light yellow sublayers, ornamented, with evenly distributed pits, the pits are circular, ellipsoid, oblong, triangular, Y-shaped to irregular. And C3 is flexible and hyaline.

It is described by Freitas and Carrenho (2013) as follows: isolated, globose spores, soft olive green surface, opaque, ornamented with small, densely packed depressions, united in combs, C1 is opaque, brittle, light cream, with pits-like ornamentation and C2 is hyaline, thin and offset from C1, thin. Also on the website (http://www.zor.zut.edu.pl) it is described as yellowish white to pale yellow in colour, globose to subglobose, produced laterally in the neck of a sporiferous sacculus. The spore wall is made up of three layers: C1 is evanescent and hyaline, C2 is laminated and yellowish white to pale yellow, decorated with evenly distributed pits and C3 is flexible and hyaline.

Morphotype 17:

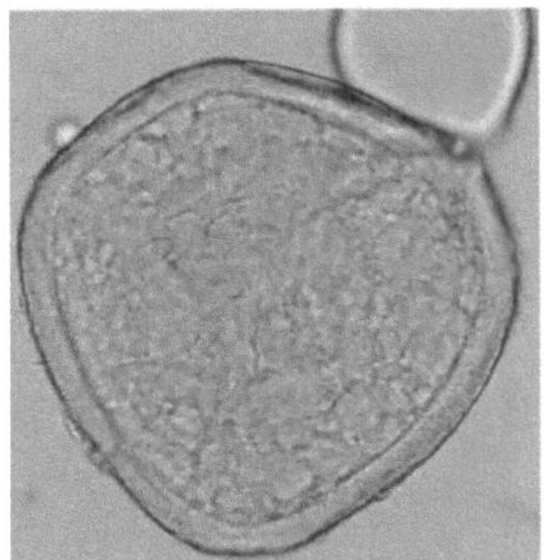

The glomerospores of this morphotype have a light yellow to greenish luminescent colour, vary in size from small to large and a very characteristic shape, mostly with a square base and pointed (pyramidal) faces. This morphotype visually resembles a species from the genus Claroideoglomus, which is described as having globose to subglobose spores, with some oblong ones; a smooth, opaque surface; a cream to ochre colour and luminescent walls. It has evanescent C1 walls, mucilaginous hyaline walls and hyaline C2 walls, which are only visible in young spores. C3 is laminated, smooth and pale yellow to ochre in colour, luminescent and C4 is smooth, flexible, apparently hyaline and thin, completely detached from C3 (http://invam.wvu.edu; FREITAS, CARRENHO, 2013).

Morphotype 18:

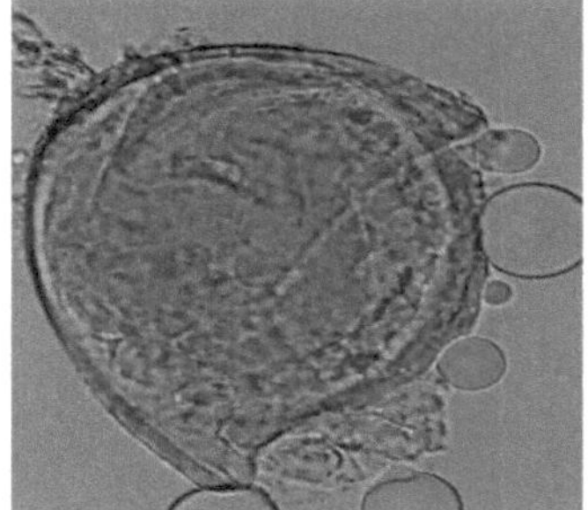

The glomerospore of this morphotype is characterised by being hyaline externally and yellow-orange internally. The size varies from small to medium and the shape is globose to subglobose, although it has a flattened appearance. The wall is made up of at least two layers. In some specimens it was possible to observe the presence of a hyaline sporogenous hypha.

This morphotype resembles the spore of Glomus pustulatum which is described (http:

isolated spores, orange in colour, globose to slightly oval in shape, presence of a single hypha, wall composed of three layers, C1 being permanent, grey-yellow to butter-yellow in colour, base circular, elliptical to irregular in plan view, unevenly distributed over the surface of the spore; C2 laminated, smooth, dark orange and C3 is flexible, hyaline and well adhered to C2.

Koske et al. (1986) describe G. pustulatum with spores formed in isolation in soil, with a pale yellow to yellow-brown colour, the presence of pustules, blister-like structures on the surface of the outer spore wall; a three-layered wall, where C1 is yellow-brown to orange-brown with a basal surface of 1-5 pm, C2 is laminated pale yellow to yellow-brown and C3 is thin, hyaline and adhered to C2.

Morphotype 19:

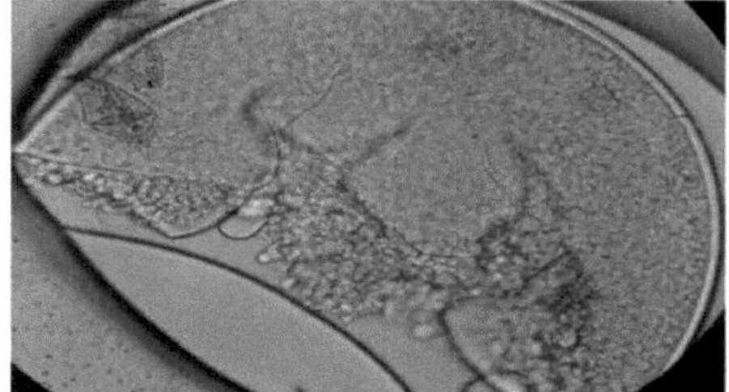

The glomerospore of this morphotype is characterised by being hyaline to white, having a globose to subglobose shape, and varying in size from small to large, although the large size is predominant.
It has a bulbous spore cell with a yellowish colour and the spore wall has at least two layers.

This morphotype resembles the Cetraspora pellucida species in that its spore is formed from a bulbous sporogenous cell, is hyaline to yellow in colour, globose to subglobose, sometimes ovoid, and is made up of a spore wall and two germinal inner walls. C1, which forms the spore surface, is permanent, smooth and hyaline and C2 is laminated and hyaline to yellow in colour (http://www.zor.zut.edu.pl).

The website (http://invam.wvu. edu) describes the spore *of C. pellucida* as hyaline to white in younger spores and yellow-brown in mature spores, globose to subglobose in shape, often elliptical or strongly oblong. The wall has three adherent layers which, in juvenile spores are of equal thickness, however the thickening of the laminated layer (C2) as the spore wall is differentiated. C1 is a permanently rigid, outer, hyaline and smooth layer that adheres to C2. C2 is hyaline to pale yellow and very thin. C3 is a flexible, hyaline and very thin layer.

Morphotype 20:

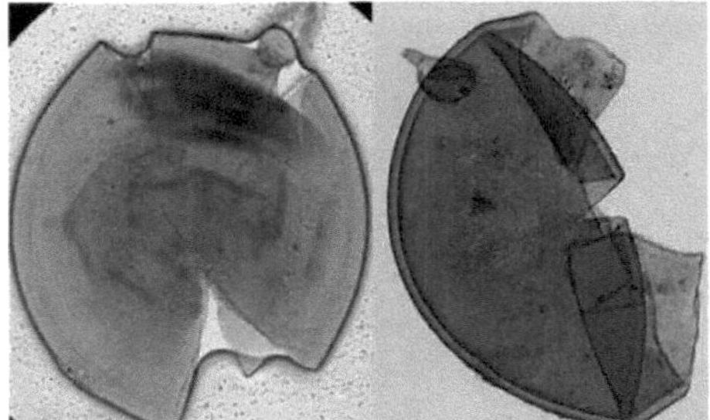

The glomerospore of this morphotype is characterised by its brownish orange to brown colour, globose shape and large size, visibly standing out from several other morphotypes. It has a bulbous sporogenous cell and the spore wall is made up of at least two layers, which were clearly visible to the PVLG.

This morphotype is similar to species of the Racocetra genus, since their spores can be coloured from brown, dark copper, orange-red, reddish brown to dark brown, and are also formed from bulbous spore cells. They have walls made up of exospores and endospores, with or without ornamentation, and a globose to subglobose shape (http://invam.wvu.edu; FREITAS, CARRENHO, 2013).

Morphotype 21:

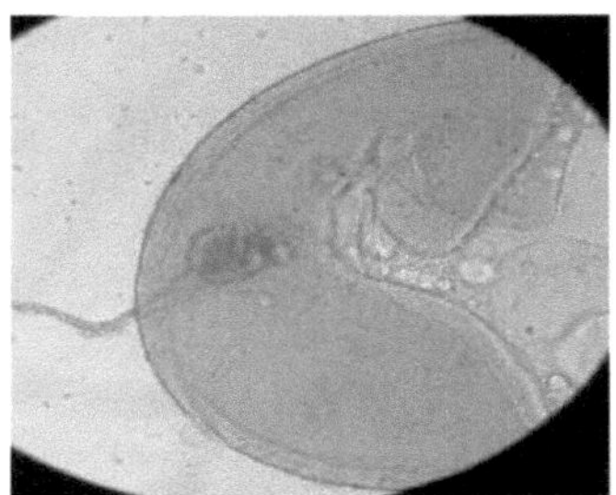

The glomerospore of this morphotype is characterised by a translucent light yellow colour, a globose to subglobose shape, a large size, the presence of a bulbous spore cell and a wall made up of unit-like layers without ornamentation.

This morphotype resembles the species of the genus Gigaspora, as its spore is characterised by being formed from a bulbous sporogenous cell, does not have ornamentation, a wall with three phenotypically different layers and auxiliary cells with equinulate projections (FREITAS, CARRENHO, 2013).

All *Gigaspora* species are known for producing spores without ornamentation, consisting of only a two-layered spore wall; germ tubes arising from a thin papillary layer (wart), originating from the inner surface of the laminate layer, as well as having thin-walled auxiliary cells with equinulate surfaces. They range in colour from white to cream, pale pink, yellowish brown, bright yellow-green to dark yellow and have a globose to subglobose and sometimes irregular shape (http://invam.wvu.edu).

Morphotype 22:

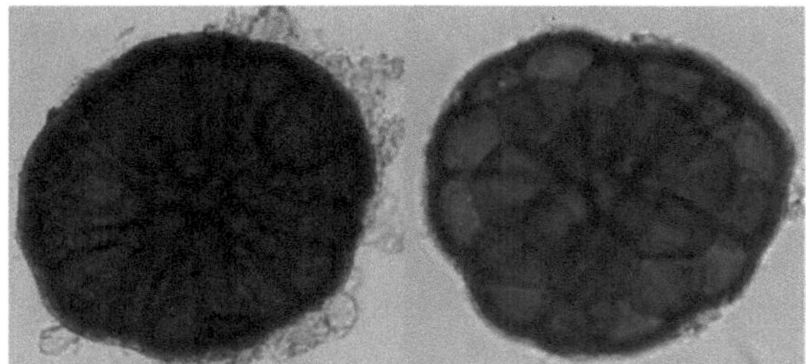

The glomerospore of this morphotype is characterised by its reddish brown colour, ornate appearance due to an aggregate between them, or sporocarps, globose to subglobose irregular and flattened shape, medium size and medium wall thickness.

This morphotype resembles the species Glomus clavisporum (= Sclerocystis clavispora), as its spore is characterised by being formed in compact sporocarps that occur in the soil, with an orange to brown colour, a globose to subglobose shape and no peridium. The spore, on the other hand, is coloured orange to brown, shaped like a clef to subcylindrical towards a hypha; the wall is made up of two layers, where C1 is evanescent and hyaline and forms the surface of the spore, rarely present in mature spores; C2 is laminated and coloured orange to brown (http://www.zor.zut.edu.pl).

Morphotype 23:

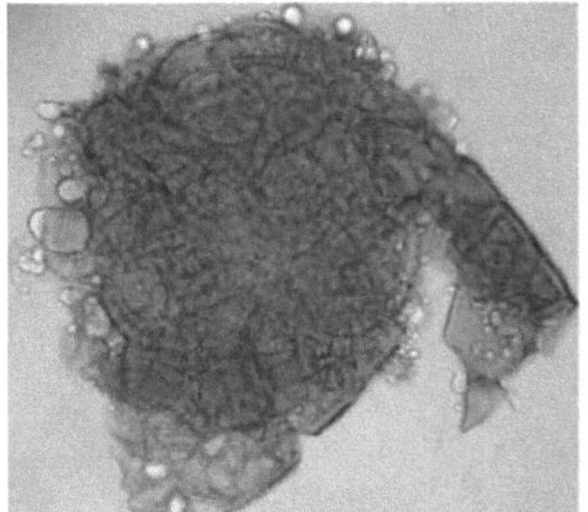

The glomerospore of this morphotype is characterised by its surface having a granular/irregular appearance when viewed under the stereomicroscope, it has a brownish colour, a subglobose shape and is large in size. Under the microscope, it looks like a tissue with several hexagon-shaped cells, and its walls have at least two layers.

This morphotype bears mere resemblance to the *Glomus sinuosum/Sclerocystis sinuosa* species when comparing the image seen on the website (http://invam.wvu.edu/the-fungi/classification/glomaceae/rhizophagus/sinuosum) to the image seen under a stereomicroscope; however, no good quality image could be acquired. However, the description given on the aforementioned website has some similar characteristics, such as: orange colour when immature, turning orange-brown to dark orange-brown when mature, globose, subglobose, occasionally pulviniform shape and irregular surface due to protruding spores covered by a dense peridium.

Morphotype 24:

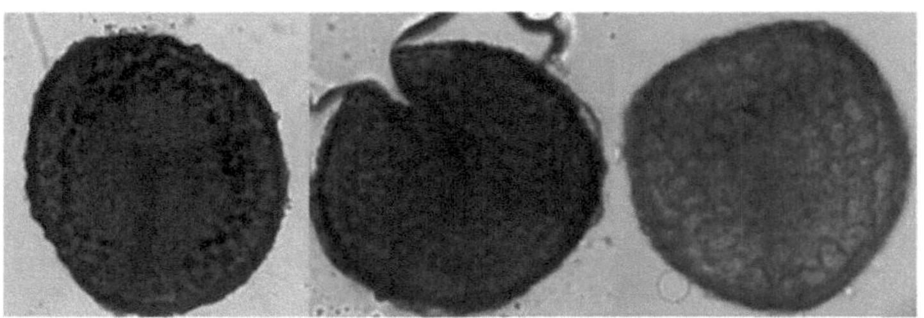

This morphotype is characterised by its light to dark brown colouration, small to medium size and rather peculiar shape, with a rounded/convex base and a pointed surface, i.e. similar to a three-sided pyramid. Its wall is ornamented and quite rigid, as well as having an outer layer similar to the evanescent layer present in some glomerospores.

When checking specific bibliographies, it was not possible to find a similar glomerospore, so it cannot be said that this type is a Glomeromycota spore. However, Pedrão and Barrilari (2001), (http://www.phoenix.org.br/Phoenix30_Jun01.htm) describe in schematic images (fig. 08) a spore fossil very similar to this type.Palynology studies microfossils of an organic nature, known as palynomorphs. Examples of palynomorphs include pollen grains, spores, acritarchs, prasinophytes, dinoflagellates, chitinozoans, scolecodonts, conchostracans, palynophoraminifera, which are found in almost the entire geological column, from the Precambrian to the present day and are preserved in different depositional environments (PEDRÃO, BARRILARI, 2001).

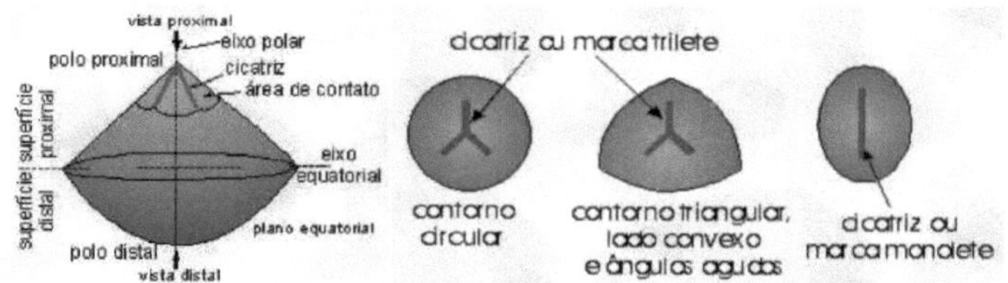

Figure 08 - Schematic diagram of the spore fossil in lateral and polar view.
Source: Pedrão and Barrilari (2001)

From the morphotypic description it was possible to give a direction to a probable classification at genus or even species level for some morphotypes, thus six probable species of the genus *Glomus* were found, four of *Acaulospora,* two of *Claroideoglomus,* the others had only one representative, *Pacispora, Funneliformis, Diversispora, Sclerocystis, Archaeospora,*

Rhizophagus, Cetraspora, Racocetra and Gigaspora. These are found in six families (Glomeraceae, Pacisporaceae, Acaulosporaceae, Diversisporaceae, Gigasporaceae and Claroideoglomeraceae).

The fact that *Gigaspora* had only one representative morphotype and appeared in only two areas may be related to the fact that species of this genus have poor adaptation to monoculture areas, since it only appeared in forage palm cultivation and in the managed Caatinga, a natural ecosystem where it appears predominantly (SILVA, 2010). It may also be related to the fact that it occurs and sporulates in soils with low fertility, a pH of less than 6 and low levels of P and organic matter (SIQUEIRA *et al.,* 1989), since the areas in which it was found had the lowest pH values, one of them had one of the lowest P values and the other had the lowest organic matter content.

CHAPTER 6

FINAL CONSIDERATIONS

The cultivation areas surveyed, despite being small and mostly used for monocultivation, had a very significant number of AMF morphotypes, where 24 different morphotypes were diagnosed, ten of which were present in all the areas, which may be related to crop rotation, since in these areas other crops (broad beans, beans, maize, fodder palm) were planted in previous years, making them generalists.

The presence and quantity of spores of some morphotypes in some areas indicates their preference for a particular type of crop. Morphotypes 4 and 14, for example, are only present in the area cultivated with maize.

Morphotypes 21 and 23 are only present in the forage palm cultivation area and in the managed Caatinga, while morphotypes 15 and 18 have a preference for the areas cultivated in consortium and forage palm and morphotype 19 is present in the bean cultivation area and in the three-crop consortium.

The highest density of spores was found in the areas under agricultural cultivation as expected, where the areas had from 399 (area 04 - cultivated by forage palm) to 985 spores (area 01 - cultivated by maize).

The managed Caatinga area, on the other hand, had only 360 spores. Therefore, all the areas had a relatively large number of spores, a fact related to the low value of assimilable P and the conditions of water stress in the soil that lead to sporulation.

Therefore, studying Arbuscular Mycorrhizal Fungi, emphasising their diversity, population and community, is a fundamentally important step for

different approaches, either in contributing to the understanding of the symbiotic role in different ecosystems, or in achieving improvements in the production of agricultural ecosystems.

BIBLIOGRAPHICAL REFERENCES

ABBRUZZINI, T. F. **Quality and quantity of soil organic matter in conventional and organic sugarcane cultivation**. 2011. 92 p. dissertation (Master of Science) - University of São Paulo, Piracicaba, 2011.

ANDREOLA, F.; FERNANDES, S. A. P. The soil microbiota in organic agriculture and crop management. In. SILVEIRA, Adriana Parada Dias; FREITAS, Sueli dos Santos. **Soil microbiota and environmental quality**. Campinas: Agronomic Institute, 2007. p. 21-33.

ALBURQUERQUE, P. P. **Diversity of Glomeromycetes and microbial activity in soils under native vegetation in the semi-arid region of Pernambuco**. 2008.113f. Thesis (Doctorate in Fungal Biology) - Federal University of Pernambuco, Recife, PE, 2008.

BAGO, B.; PFEFFER, P.E.; SHACHAR- HILL, Y. Carbon metabolism and transport in arbuscularmycorrhizas. **Plant Physiol.**, v.124, p.949-957, 2000.

BARROS, A. T. **Physico-chemical and biological characterisation of water and soils From the banks of the Piancó River, PB.** 2015. 192f. Thesis (Doctorate in Materials Science and Technology). Federal University of Campina Grande, Campina Grande, PB, 2015.

BAUMGARTNER, K.; SMITH, R. F.; BETTIGA, L. Weed control and cover crop management affect mycorrhizal colonisation of grapevine roots and arbuscular mycorrhizal fungal spore populations in a California vineyard. **Mycorrhiza,** v.15, n.2, p.111-119, 2005.

BERBARA, R. L. L.; SOUZA, F. A.; FONSECA, H. M. A. C. Arbuscular mycorrhizal fungi: far beyond nutrition. In: FERNANDES, M.S. (Ed.). **Mineral Nutrition of Plants**. Brazilian Society of Soil Sciences, Viçosa, 2006, 432p.

BERCH, S. M.; KOSKE, R. E. *Glomuspansihalos: A* new species in the Endogonaceae, Zygomycetes. **Mycologia**, v.78, n. 5, p.832-836, 1986.

BORBA, M. F.; AMORIM, S. M. C. Arbuscular mycorrhizal fungi in evergreens: a subsidy for cultivation and replanting in degraded areas. **Revista de Biologia e Ciências da Terra**, v.7, n.2, p.20-27, 2007.

BRANDÃO, E. M. The components of the soil microbial community. In: CARDOSO, E. J. B. N.; TSAI, S. M. NEVES, M. C. P., (Cords) **Microbiologia do solo**. Campinas: Brazilian Society of Soil Science, 1992. p.1-15.
CARDOSO, E. J. B. N. CARDOSO, I. M., NOGUEIRA, M. A., BARRETA, C. R. D. M., PAULA, A. M. Arbuscular mycorrhizae in the acquisition of nutrients by plants. In: Siqueira J.O.;

SOUZA.FA.; CARDOSO E. J.B.N.; TSAI, S.M. (Eds) **Mycorrhizae:** 30 years of research in Brazil. Lavras:

UFLA, 2010, Chap. 6, p. 153-195.

CARRENHO, R., GOMES-DA-COSTA, S. M., BALOTA, E. L., COLOZZI-FILHO, A. Arbuscular mycorrhizal fungi in Brazilian agrosystems. In: SIQUEIRA, J.O.; SOUZA, F.A.; CARDOSO E.J.B.N.; TSAI, S.M. (Eds) **Mycorrhizae: 30 years of research in Brazil**. Lavras: UFLA, 2010, Chap 7, p. 215-249.

COELHO, R. A. **Mycorrhizal colonisation, nutrition and morphology of coffee trees in monoculture and agroforestry systems.** 2008. 83 f. Dissertation (Master's in Agronomy) - State University of Southwest Bahia, Vitória da Conquista, 2008.

COLOZZI-FILHO, A.; CARDOSO, E. J. B. N. Detection of arbuscular mycorrhizal fungi in roots of coffee trees and crotalaria grown between the rows. **Pesquisa Agropecuária Brasileira**, Brasília, v.35, n.10, p.2033-2042, 2000.

COLOZZI-FILHO, A; NOGUEIRA, M. A. **Arbuscular Mycorrhizae in Tropical Plants:** Coffee, Cassava and Sugarcane. Ed. A. Silveira and Sueli Freitas. Campinas: Agronomic Institute, 2007.

COSTA, C.M.C. et al. Influence of arbuscular mycorrhizal fungi on the growth of two genotypes of witch hazel (Malgiphiaemarginata D.C.). **Pesquisa Agropecuária Brasileira**, v.35, n.4, p.893-901, 2001.

DE SOUZA, F. A.; DA SILVA, I. C. L.; BERBARA, R. L. L. Arbuscular mycorrhizal fungi: much more diverse than imagined. In: MOREIRA, F.M.S.; SIQUEIRA, J.O.; BRUSSAARD, L. (Eds). **Soil biodiversity in Brazilian ecosystems**. Lavras: UFLA, 2008. p.482-536.

EMBRAPA. National Soil Research Centre (Rio de Janeiro, RJ). **Brazilian soil classification system.** Brasília: Embrapa Produção de Informação; Rio de Janeiro: Embrapa Solos, 1999. V.26, 412p. il. CDD 631.44.

FERNANDES, A. B., SIQUEIRA, J. O., MENEZES, M. A. L., GUEDES, G. A. A. Differential effect of P on the establishment and effectiveness of endomycorrhizal symbiosis in maize and soya. **Revista Brasileira Ciência do Solo**, Campinas, v.11, n.2, p.101-108, 1987.

FERNANDES, S. G., MACHADO, C. T. I., IGNÁCIO, I. G., PEREIRA, C. D., FERNANDES, L. A. **Mycorrhizal Activity in Areas of Family Farmers in Northern Minas Gerais.** In: XXXI REUNIÃO BRASILEIRA DE FERTILIDADE DO SOLO, FERTEBIO, 2012, Macéio - Al.

FOLLI-PEREIRA, M. S., MEIRA-HADDAD, L. S., BAZZOLLI, D. M. S., KASUYA, M.C.M. Arbuscular mycorrhiza and plant tolerance to stress. **Revista Brasileira Ciência do Solo**, v.36, p.1663-1679. 2012.

FREITAS, R. O.; CARRENHO, R. **Digital guide to arbuscular mycorrhizal fungi from the Adolpho Ducke Forest Reserve and PDBFF Reserve**, Manaus, 2013. 52p. Available at: < https://ppbio.inpa.gov.br/sites/default/files/Guia%20de%20FMA Vers%C3%A3o%201.pdf >. Accessed on: 12 September 2014.

GERDEMANN, J. W; NICOLSON, T. H. Spores of mycorrhizalEndogene species extracted from soil by wet sieving and decanting. **Transactions of the British Mycological Society,** 46, p.235- 244, 1963.

GOTO, B. T.; MAIA, L. C. Glomerospores: a new name for spores of Glomeromycota, a molecularly distinct group of Zygomycota. **Mycotaxon**, v.96, p.120-132, 2006.

GUSMÃO, L. P.; MAIA, L. C. Fungi. In: GUSMÃO, L. P; MAIA, L.C. (Eds). **Diversity and characterisation of fungi from the Brazilian semi-arid** region. Recife, Associação de Plantas do Nordeste - APNE, 2006, v.2, p.27-47.

GUSMÃO, L. F. P.; MELO, E.; FRANÇA, F.; BARBOSA, F. F. The Brazilian semi-arid region and fungi. In: GUSMÃO, L. F. P.; MAIA, L. C. (Eds.). **Diversity and characterisation of fungi from the Brazilian semi-arid** region. Instituto Milênio do Semiárido - IMSEAR, 2006, v.2, p.19-26.

HOFFMAN, L. V.; LUCENA, V. S. **Para entender micorrizas arbusculares.** Embrapa Cotton (documents 156), 2006.

BRAZILIAN INSTITUTE OF GEOGRAPHY AND STATISTICS. Boa Vista - PB, IBGE 2010. Available at: http://www.cidades.ibge.gov.br/painel/painel.php?lang=&codmun=250215&search=%7Cboa-vista. Accessed on: 22 Mar. 2014.

JAMES, T.Y., KAUFF, F., SCHOCH, C. L., MATHENY, P. B. Reconstructing the early evolution of fungi using a six-gene phylogeny. **Nature**, v.443, p. 818-822, 2006.

KELLY, R.M.; EDWARDS, D.G.; THOMPSON, J.P.; MARGAREY, R.C. Growth responses of sugarcane to mycorrhizal spore density and phosphorus rate. Australian **Journal of Agricultural Research**, v.56, p.1405-1413, 2005.

KOSKE, R. E. FRIESE, C., WALKER, C, DALPÉ, Y. Glomus pustulatum: a new species in the Endogonaceae. **Mycotaxon**, v. 26, p.143-149, 1986.

LAMBAIS, M.R; RAMOS, A.C. Signalling and signal transduction in arbuscular mycorrhizae. In: SIQUEIRA J.O., SOUZA, F.A.,CARDOSO, E.J.B.N.; TSAI, S.M. (Eds) **Mycorrhizae:** 30 years of research in Brazil. Lavras: UFLA, 2010, Chap. 4, p.118-132.

LEAL, I. R., TABARELLI, M., SILVA, J. M. C. (Eds.) **Ecologia e conservação da Caatinga.** Recife: University (Federal University of Pernambuco), 2003. 822p.

MAIA, L.C., SILVA, G. A., YANO-MELO, A. M., GOTO, B. T. Arbuscular Mycorrhizal Fungi in the Caatinga Biome. In: SIQUEIRA, J.O., SOUZA, F.A., CARDOSO E.J.B.N., TSAI, S.M. (Eds) **Mycorrhizae:** 30 years of research in Brazil. Lavras: UFLA, 2010. Chap. 10, p. 311-335.

MALAVOLTA, E. **ABC da Adubação.** 4 ed. São Paulo: Agronômica Ceres, 1979. 189p.

MARIANO, L. P. **Fertility assessment of soils used for coffee farming by family producers in Nova Resende**. 2010. 33f. Final Course Work (Technology in Coffee Growing) Federal Institute of Education, Science and
Tecnologia do Sul de Minas Gerais, Muzambinho, 2010.

MARSCHNER, H.; DELL, B. Nutrient uptake in mycorrhizal symbiosis. **Plant Soil**, v.159, p.89- 102, 1994.

MEDEIROS, J. C., ALBUQUERQUE, J. A., MAFRA, A. L., ROSA, J. D., GATIBONI, L. C. Calcium:magnesium ratio of soil acidity corrector in
nutrition and the initial development of maize plants in a Cambissolo.
Semina: Ciências Agrárias, Londrina, v. 29, n. 4, p. 799-806, 2008.

MELO, A. S. T., RODRIGUEZ, J. L. **Paraíba: economic development and the environmental question.** João Pessoa: Grafset, 2004.208p.

MELLO, C. M. A., SILVA. I. R., PONTES, J. S., GOTO, B. T., SILVA, G. A., MAIA, L. C. Diversity of arbuscular mycorrhizal fungi in a Caatinga area, PE, Brazil. **Acta Botânica Brasílica**, v.26, n.4, p.938-943, 2012.

MERGULHÃO, A. C. E. S, FIGUEIREDO, M. V. B , OLIVEIRA, J. P., SILVA, M. L. R. B., BURITY, H. A. **Mycorrhizae (arbuscular mycorrhizal fungi)** - biological input for use in agriculture. Agronomic Institute of Pernambuco: Government of the State of Pernambuco. 2008. Available at:< http://www.ipa.br/resp61.php> Accessed on: 15 Sep. 2015.

MERGULHÃO, A. C. E. S., FIGUEIREDO, M. V. B., BURITY, H. A., MAIA, L. C. Hosts and successive multiplication cycles affect the detection of arbuscular mycorrhizal fungi in areas impacted by gypsum mining. **Revista Árvore**, Viçosa, v.33, n.2, p.227-236, 2009.

MINISTRY OF THE ENVIRONMENT (MMA). Caatinga. Available at:< http://www.mma.gov.br/biomas/caatinga> Accessed on: 08 Oct. 2014.

MIRANDA, J. C. C.; MIRANDA, L. N. Arbuscular mycorrhiza. In: VARGAS, M. A. T.; HUNGRIA, M., (Ed.). **Biology of cerrado soils**. Planaltina: EMBRAPA-CPAC, 1997. p. 67-111.

MIRANDA, J. C.C; VILELA, L; MIRANDA, L. N. Dynamics and contribution of arbuscular mycorrhiza in a production system with crop rotation. **Pesquisa Agropecuária Brasileira**, v. 40, n.10, p.1005-1014, 2005.

MIRANDA, J. C. C.; MIRANDA, L. N. **Impacto do sistema direto na diversidade de fungos micorrízicos arbusculares nativos em solo de Cerrado.** Planaltina/ DF: Embrapa Cerrados, 2007 (Embrapa Cerrados, TECHNICAL COMMUNICATION, 135).

MOHAMMAD, A.; MITRA, B.; KHAN, A.G. Effects of sheared-root inoculum of Glomus Intraradices on wheat grown at different phosphorus levels in the field. **Agriculture Ecosystem and Environment**, v.103, p.245-249, 2004.

MOREIRA, F. M. S.; SIQUEIRA, J.O. **Microbiologia e Bioquímica do solo**. 2.ed. Lavras: UFLA, 2006.729p.

OEHL, F.; SIEVERDING, E. Pacispora, a new vesicular arubscular mycorrhizal fungal genus in the *Glomeromycetes*. **Journal of Applied Botany** , v.78, p.72-82, 2004.

OLSSON, P. A.; THINGSTRUP, I.; JAKOBSEN, I.; BAATH, F. Estimation of the biomass of arbuscularmycorrhizal fungi in a linseed field. **Soil Biol. Biochem.**, v.31, p.1879-1887, 1999.

ORLANDO FILHO, J. et al. K, Ca, Mg relationship of quartz sand soil and sugarcane productivity. **SITAB**, Piracicaba, v.14, n. 5, p. 13-17, 1996.

PEDRÃO, E.; BARRILARI, I. M. R. **The palynomorphs:** spores. Phoenix, v.3, n. 30, 2001. Available at:< http://www.phoenix.org.br/Phoenix30 Jun01.htm>. Accessed on: 25 Feb. 2015.

PRADO, D. The Caatingas of South America. In: LEAL, I. R., TABARELLI, M., SILVA, J. M. C. (Eds.) **Ecologia e conservação da Caatinga**. Recife: Ed.Universitária (Federal University of Pernambuco), 2003, p. 3-73.

PRIMAVESI, A. **Manejo ecológico do solo:** a agricultura em regiões tropicais. São Paulo: Nobel, 2002.

REDECKER, D.; MORTON, J.B.; BUNS, T.D. Ancestral lineages of arbuscular mycorrhizal fungi (Glomales). **Mol. Phylogenet. Evol.**, v.14, p.276-284, 2000.

REDECKER, D. SCHÚBLER, A., STOCKINGER, H., STÚRMER, S., MORTON, J., WALKER, C. An evidence-based consensus for the classification of arbuscular mycorrhizal fungi (Glomeromycota). **Mycorrhizae doi** : 10.1007/s00572-013-0486-y, 2013.

RIBEIRO, A.C.; GUIMARÃES, P.T.G.; ALVAREZ V., V.H. (ed.). **Recommendations for the use of soil improvers and fertilisers in Minas Gerais:** 5ª Approximation. Viçosa: Soil Fertility Commission of the State of Minas Gerais, 1999. 359p.

RILLIG, M. C., MURMMEY, D. L. Mycorrhizas and soil structure. **New Phytologist**, v.171, p. 4153, 2006.

RODRIGUEZ, J. L. **Atlas Escolar Paraíba:** geo-historical and cultural space. 4th edition, expanded and updated. Grafset: João Pessoa -PB, 2012. 192p.

SAGGIN-JÚNIOR, O. J. ; LOVATO, P. E. Application of arbuscular mycorrhizae in the production of seedlings and micropropagated plants. In: SIQUEIRA, J. O. (Org.). **Interrelation between fertility, soil biology and plant nutrition**. Viçosa: SBCS, Lavras: UFLA/DCS, 1999.

SAGGIN JÚNIOR, O. J. ; SILVA, E. M. R. Arbuscular Mycorrhiza - Role, Functioning and Application of Symbiosis. In: A. M. de Aquino; R. L. de Assis. (Org.). **Biological Processes of the Soil-Plant System - Tools for Sustainable Agriculture**. 1 ed. Brasília: Embrapa Agobiologia, Embrapa Informação Tecnológica, 2005, Chap. 1, p. 101-149.

SANTOS, E. S. **Caderno Pedagógico de Química:** Physical-chemical analyses of water and soil. Pinhais, SP. 2008, 61p.

SCHENCK, N.; PERÉZ, Y. **Manual for the identification of VA mycorrhizal fungi.** In: Synergistic Publications, 3ed. Gainesville, 1987. 287p.

SCHUBLER, A.; WALKER,C. **The glomeromycota:** A species list with new families and new genera. Gloucester: England, 2010.

SELOSSE, M.A.; BAUDOIN, E.; VANDENKOORNHUYSE,P. Symbiotic microorganisms, a key for ecological success and protection of plants. **CR Biol.**, v.327, p.639-648, 2004.

SENGIK, E.S. **Macronutrients and Micronutrients in Plants**. Version 2003. Available at: < www.nupel.uem.br/nutrientes-2003pdf > Accessed on: 31 Mar. 2015.

SIEVERDING, E. Function of VA mychorrhiza. In: SIEVERDING, E. (Ed). **Vesicular-arbuscular Mycorrhiza Management in Tropical Agrosystems**. Eschborn: Technical Cooperation Federal Republic of Germany, 1991, p.57-70.

SILVA, G.A., MAIA, C. L., SILVA, F. S. B., LIMA, P. C. F. Potential infectivity of arbuscular mycorrhizal fungi from native caatinga and mining degraded areas in the state of Bahia. **Revista Brasileira de Botânica**, v. 24, n. 2, p. 135-143, 2003

SIQUEIRA, J. O.; FRANCO, A. A. **Biotecnologia do solo, fundamentos e perspectivas.** Brasília: MEC/ABEAS; Lavras: ESAL/FAEPE, 1988. 236 p.

SIQUEIRA, J.O.; COLOZZI-FILHO, A.; OLIVEIRA, E. Occurrence of arbuscular versicular mycorrhizae in agro- and ecosystems of the State of Minas Gerais. **Pesquisa Agropecuária Brasileira**, v.24, n.12, p.1499-1506, 1989.

SIQUEIRA, J.O.; MOREIRA, F.M.S. **Soil microbiology and agricultural sustainability:** a focus on soil fertility and plant nutrition. In: BRAZILIAN MEETING ON SOIL FERTILITY AND PLANT NUTRITION. SBCS: MANAUS, 1996, p. 1-42.

SMITH, S. E.; READ, D. J. **Mycorrhizal symbiosis.** New York: London Academic Press, 1997.

SOUTO, P. C., SOUTO, J. S., PAES DE MIRANDA, J. R., SANTOS, R. V., ROCHA ALVES, A. Edaphic microbial community and mesofauna in soil under caatinga in the semi-arid region of Paraíba. **Rev. Bras. Ciênc. Solo**, v.32, n.1, p. 151-160, 2008.

SOUSA, R. F., BARBOSA, M. P., SOUSA JÍNIOUR, S. P., NERY, A. R., LIMA, A. N. Estudo da evolução espacial-temporal da cobertura vegetal do município de Boa Vista- PB utilizando Geoprocessamento. **Caatinga**. Mossoró, Brazil, v.21, n.3, p.22-30, 2008.

SOUZA, F. A., STUMER, S. L., CARRENHO, R., TRUFEM, S. F. B. Classification and taxonomy of arbuscular mycorrhizal fungi and their diversity and occurrence in Brazil. In: Siqueira J.O., Souza, F. A., CARDOSO, E. J. B. N.; TSAI, S. M. (Eds.) **Mycorrhizae:** 30 years of research in Brazil. Lavras: UFLA, 2010, Chap. 2, p. 15-73.

SOUZA, R. G.; MAIA, L. C.; SALES, M.F.; TRUFEM, S. F. B. Diversity and infectivity potential of arbuscular mycorrhizal fungi in a Caatinga area, Xingó Region, Alagoas State, Brazil. **Revista Brasileira de Botânica,** v.26, n.1, p.49-60, 2003.

STEVENSON, F. J. **Humus chemistry:** genesis, composition, reactions. 2. ed. New York: John Wiley & Sons, 1994.

STÚMER, S.L; SIQUEIRA, J. O. **Diversity of Arbuscular Mycorrhizal Fungi in Brazilian Ecosystems.** In: MOREIRA, F. M. S.; SIQUEIRA, J. O.; BRUSSAARD, L. (Eds). Soil Biodiversity in Brazilian Ecosystems. Lavras: UFLA, 2008, p.537-583.

YANO-MELO, A. M., GOTO, B. T., SILVA, G. A., MAIA. L. C. Diversity of arbuscular mycorrhizal fungi in Brazil. In: JARDIM, M.A.G.; BASTOS, M.N.C.; SANTOS, J.U.M. (Eds). **Challenges for Brazilian Botany in the New Millennium:** inventory, systematisation and

conservation of plant diversity. CONGRESSO NACIONAL DE BOTÂNICA, Belém: MPEG/UFRA/EMBRAPA, 2003, p.173-174.

TOMÉ JR., J. B. **Manual for interpreting soil analyses**. Guaíba: Agropecuária, 1997. 246p.

Printed by Books on Demand GmbH, Norderstedt / Germany